Architected Metamaterials: Design Principles and Properties

Nikhil Gupta • Caleb Beckwith

Architected Metamaterials: Design Principles and Properties

Springer

Nikhil Gupta
Mechanical and Aerospace Engineering
Department, Tandon School of Engineering
New York University
Brooklyn, NY, USA

Caleb Beckwith
Mechanical and Aerospace Engineering
Department, Tandon School of Engineering
New York University
Brooklyn, NY, USA

ISBN 978-3-031-98329-0 ISBN 978-3-031-98330-6 (eBook)
https://doi.org/10.1007/978-3-031-98330-6

This Springer imprint is published by the registered company Springer Nature Switzerland AG
The registered company address is: Gewerbestrasse 11, 6330 Cham, Switzerland

Acknowledgments

The authors thank the National Science Foundation grant CMMI-2036802 through the Future Manufacturing program, the American Society for Non-destructive Testing Fellowship Award 2024, and the NYU Center for Advanced Technology in Telecommunications (CATT) grant in 2023–2024 and 2024–2025 for supporting the study. The views presented are from the authors, not from the funding agencies.

The authors thank Guan-Lin Chen for helping with the manuscript preparation. The authors also acknowledge their families for their support throughout the development of this work.

Contents

Chapter 1
Metamaterials: Introduction

1.1 Introduction

Metamaterials (MMs) is a class of engineered materials designed for specific functionalities, ranging from the manipulation of electromagnetic and acoustic waves to more exotic properties such as negative thermal expansion [1–3]. These properties are generated by combining a specific material type with micro- and macro-structures that can manipulate the stimulus such as light waves or mechanical loads in ways natural materials are unable to do. Metamaterials are at the forefront of technological advancements, enabling functionalities such as optical cloaking, and energy harvesting [4–6]. Once mostly a subject of theoretical interest, the advanced manufacturing techniques have helped in manufacturing novel micro- or macro-structures of metamaterials with predetermined and tuned material properties to deliver the desired functionality [7, 8], which has resulted in significant research activity in this field in recent years.

The artificial structures engineered for these materials exhibit properties not found naturally in conventional materials. Unlike these conventional materials, which derive their properties from chemical composition, metamaterials obtain their unique behaviors from the geometric arrangement of their internal strictures. These structures, typically in the form of a lattice whose length is a fraction of the manipulated phenomena, allow them to interact with electromagnetic, acoustic, thermal, or mechanical waves in novel ways. The term "meta" meaning "beyond", implies the ability of these materials to excel past the limitations of conventional materials. A foundational principle of metamaterials is their macroscopic properties, such as refractive index, density, stiffness, and/or permittivity are governed by their shape, orientation, and spatial distribution of unicells. These unit cells act as the meta-atom governing the overall shape and performance of these materials. Engineers can tailor the material's response to external stimuli, in the form of waves, force, and heat transfer, by manipulating the shape and composition of these

N. Gupta, C. Beckwith, *Architected Metamaterials: Design Principles and Properties*, https://doi.org/10.1007/978-3-031-98330-6_1

unicells. This principle enables the realization of exotic effects such as negative refractive index, band gaps, cloaking, and programmable mechanical behavior.

Through this understanding of metamaterials, heterogeneous structures can be modeled as continuous mediums with average material parameters. In electromagnetic metamaterials, structures smaller than the wavelength of light can produce negative refractive indices by simultaneously exhibiting negative electric permittivity and magnetic permeability. Similarly, acoustic metamaterials can achieve negative bulk moduli, allowing for sound insulation, cloaking, or lensing beyond what natural materials can provide. In mechanical metamaterials, architected microstructures can be used to control stress distributions, mechanical anisotropy, or even auxetic behavior, wherein the material expands laterally when stretched. A key design feature of metamaterials is their lattice structure, often inspired by crystallography but constrained by natural symmetrical limitations.

There are many possible ways of classifying metamaterials. The classifications can be developed based on the functionality, microstructure, base material type, microstructure architecture, or the application. In the present case, metamaterials are classified into four main categories, each defined by their interaction with specific types of waves and stimuli: electromagnetic metamaterials, acoustic metamaterials, thermal metamaterials, and mechanical metamaterials. A review of more than 390 scholarly articles published in various scientific journals over the past 10 years helped in developing Table 1.1, where these materials are classified based on their functionality, applications, material types, and material properties. The studies are focused on using analytical methods to develop microstructures for the desired functionality. The development process of these materials typically follows a rigorous design cycle, starting with the desired functionalities and properties as well as boundary conditions to flush out the theory behind the material structure. This is followed by extensive numerical modeling via methods such as finite element analysis (FEA) to determine the optimal composition and geometry. Once simulations validate the theoretical predictions, the metamaterial is manufactured, and its performance is evaluated [9–12]. Many of the metamaterial structures work in a narrow set of parameters such as specific optical wavelength or temperature ranges. Their performance may be severely limited or deteriorate outside of the conditions they are designed for.

Advances in manufacturing, particularly in additive manufacturing (AM) techniques, have been crucial in prototyping and manufacturing the complex designs in metamaterials [13, 14]. AM technologies such as material extrusion, vat polymerization, and powder bed fusion (PBF) have been widely used based on the specific geometrical features of the metamaterial design and the capabilities of the manufacturing method such as material type and resolution for producing these sophisticated geometries. These methods provide the precision and flexibility needed to realize the intricacies of MMs.

Figure 1.1 graphically represents various classes of metamaterials and highlights the key areas of interest in metamaterials research, including energy absorption, elastic behaviors, and cloaking mechanisms. This figure illustrates the functionalities and potential applications of each class of material. The foundational principles

Table 1.1 Distribution of papers surveyed with relevant compositions and evaluated material properties

Metamaterial class	Functionality	Application	Material type	Relevant material properties
Mechanical metamaterials	Mechanical absorption / auxetic materials / programmable stiffness	Light weight structures / prosthetic implants / robotic manipulators / vibration damping / seismic protection / automotive crash protection	Aluminum / steel / rubber / ABS / PLA / PETG / TPU / silicone / silicon / invar / titanium / PEEK / copper / polyamide / PET / ceramic foam / iron-invar / carbon fiber / PVA / graphene / tungsten / oxide / PET / tin oxide / PE / vanadium dioxide / silver	Elastic modulus / stiffness / damping / negative density / mass / gain / transmission / volume / thermal expansion coefficient / Poisson's ratio / absorption / conductivity / emissivity
Thermal metamaterials	Thermal cloaking / thermal absorption / negative thermal expansion	Heat flux control / thermal insulation / thermoelectric devices / solar energy harvesting / temperature control / infrared emission control	Vanadium dioxide / polycarbonate / TPU / nylon / aluminum / copper / UV curing resin / silicon / silicone / steel / graphene / PTFE / ABS / PLA / PEEK / gallium / gold / titanium / polyurethane / silver / PTFE / silver / Lead	Absorption / density / elastic modulus / permeability / stiffness / Impedance / permittivity / conductivity / emissivity / heat capacity / melting point
Electromagnetic metamaterials	Electromagnetic cloaking / electromagnetic absorption / nonlinear optical manipulation	Meta-Antenna / detection systems / radio frequency harvesting / cloaking devices / medical imaging / chemical sensors	Graphene / copper / silicon / aluminum / gold / graphene / PTFE / silver / vanadium dioxide / polyamide / polycarbonate / rubber / steel / tungsten / nylon / UV curing resin / PTFE	Absorption / electric shielding effectiveness / Fermi energy / radiative transfer / refractive index / transmission / magnetic field strength / permeability / permittivity

(continued)

Table 1.1 (continued)

Metamaterial class	Functionality	Application	Material type	Relevant material properties
Acoustic metamaterials	Acoustic cloaking / acoustic absorption / adaptive sound attenuation	Asymmetric transmission / underwater sonar system / noise—Cancelling devises / isolation systems / waveguides / vibration dampening	PLA / ABS / TPU / UV curing resin / polyurethane / PVC / acrylic / aluminum / porcelain / glass / silicon nitride	Refractive index / permeability / permittivity / sound transmission loss / attenuation / Impedance / transmission coefficient / sound absorption coefficient

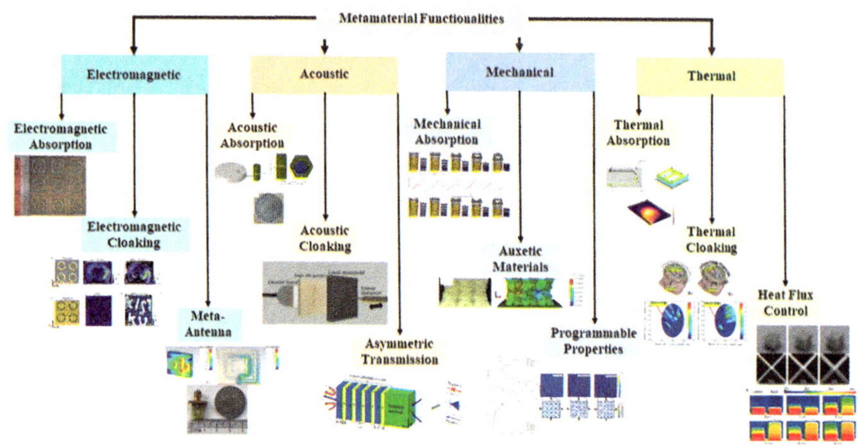

Fig. 1.1 Metamaterials are categorized by the properties they are designed for and further disambiguated by their functionalities. Electromagnetic metamaterials function as cloaking materials with one of their applications being used as an antenna. Acoustic metamaterials similarly can absorb and cloak acoustic signals and can be applied as an asymmetric acoustic diode. Mechanical metamaterials can be made with auxetic properties, and mechanically absorb forces, and can be developed with tailored properties to fit specific applications. Thermal metamaterials function to absorb and cloak against heat using conduction, convection, and radiative techniques, and are applied in applications where controlling the fluctuations in heat are important. (From left to right the pictures provided have been adapted from [15–26])

and manufacturing processes of current metamaterials will be discussed in detail in later chapters and the existing challenges will be discussed, that, once overcome, could facilitate their broader applications beyond laboratory settings. Metamaterials classified as four main categories present overlap in their functionalities. Generally, they are designed for achieving manipulation of physical or mechanical properties, manipulation of kinetic, thermal, and electromagnetic energy, and producing a cloaking phenomenon for optical, thermal, or electromagnetic waves. There may be an overlap in the properties of these materials as the focus is increasing on designing multifunctional materials.

1.2 Archetype of Functionality

The architecture of metamaterials is meticulously aligned with their specific intended uses, leading to a diverse array of structural configurations, from hexagonal lattice formations to more complex arrangements of concentric ring resonators and cascading maze-like patterns [27, 28]. These custom structures are pivotal in selectively capturing and manipulating waves across specific frequencies, and other stimulus to achieve desired functionalities.

Electromagnetic MMs and acoustic MMs are designed for precise control of electromagnetic and acoustic waves spanning a wide frequency spectrum from the lower kilohertz (kHz) to the terahertz (THz) ranges. This control is crucial in applications such as secure communications, invisibility to sonar and advanced sensing technologies. Figure 1.2 illustrates the range of frequencies observed in acoustics and electromagnetics highlighting notable frequency bands studied for MM development. Mechanical MMs and thermal MMs are engineered to modulate the transmission of kinetic, potential, and thermal energy catering to applications that require materials with specialized thermal and mechanical properties.

The first large category of metamaterials is related to the manipulation of physical properties. Materials composed of 2D and 3D lattice structures whose elastic modulus, stiffness, and Poisson's ratio are all programmed such that the material performs according to the expectation of the requested application. Advancements

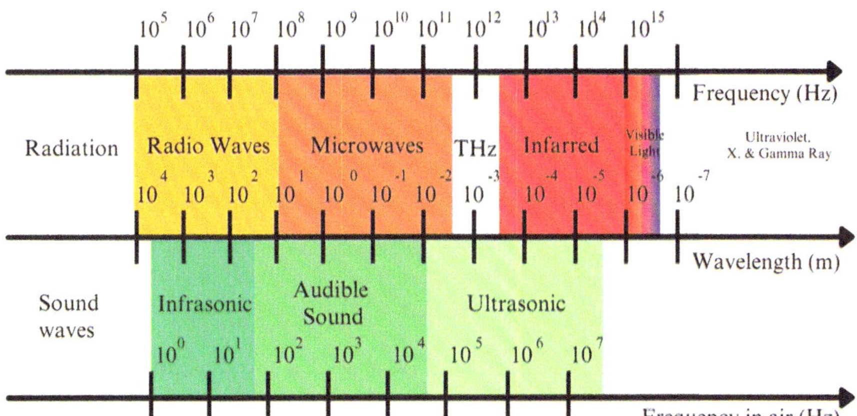

Fig. 1.2 Frequency range for acoustic sound waves as well as radiation and light waves as a spectrum

in computational design, such as machine learning and physics informed topology optimization, have been instrumental in tailoring the mechanical properties of MMs at the unit cell level to meet specific applications and design requirements [29, 30]. These technologies are expected to help in further refining the design of metamaterials and developing designs of metamaterials for new functionalities as the new algorithms are developed and the technology matures.

Often metamaterials result in mechanical responses that can be seen as unfavorable or outside what is considered natural. Anisotropy is a common phenomenon in composite materials where the material might show strength in tension but in torsion or compression the material fails easily [31, 32]. For metamaterials this is often the case as their geometry results in a skewed or directional response. Anisotropy is the opposite of isotropy where the material exhibits the same response regardless of where the force is applied. Alternatively, another common mechanical response results in the material having a negative Poisson's ratio [33]. This is also referred to as having auxetic material quality. Whereas the material is pulled in tension it expands and contracts when compressed. Giving a material a negative Poisson's ratio is often done intentionally in metamaterial design as controlling Poisson's ratio is useful in the development of compliant materials and programmable dynamic response for shock loading. Figure 1.3 provides two examples of material responses that result from the metamaterial design. Figure 1.3a shows an anisotropic response in a material where the same force applied along different parts of a material illicit a different response. This occurs in all forms of metamaterials where their design often poses some directionality, and if the simulating phenomenon is not along the incident plane the metamaterial either will not respond or the response

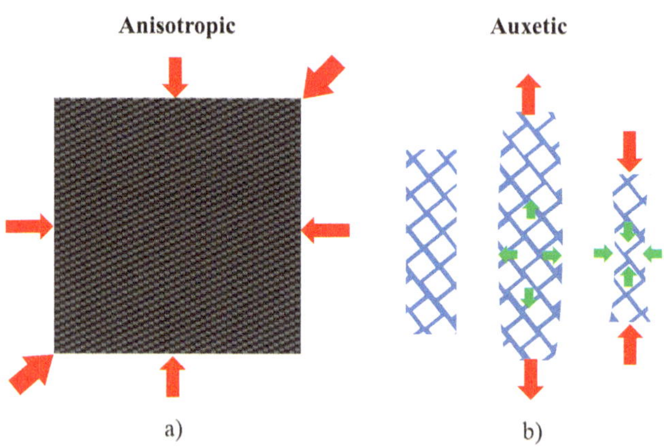

Fig. 1.3 Two outcomes of mechanical metamaterial design. (**a**) Anisotropic behavior of material where force applied along different parts of the material results in different responses from the material rather than isotropic behavior where the response would be the same regardless of location along with the material. (**b**) Auxetic behavior where the material in tension or compression exhibits a negative Poisson's ratio response where the material expands in tension and retracts in compression where typical materials with positive Poisson's ratios exhibit the opposite response

will be greatly diminished from proposed. Figure 1.3b shows an auxetic response whereas the proposed material is subject to tension or compression results in a negative Poisson's ratio response. This means that in tension the material expands, and in compression the material contracts.

Metamaterials that are focused on energy manipulation, are designed to absorb, convert, or store energy from incident stimuli [34, 35]. There are ways this controlled and directed energy-material interaction is accomplished. Deformations in the material are often caused by forces or vibrations producing strains capable of being converted to electricity through piezoelectric effect. Radiation and heat are often converted to electricity through thermal photovoltaics. This has opened new opportunities in communications for designing multiband receiving and communicating antennas and phononic diodes transmitting signals in a singular direction. Figure 1.4 depicts a band gap graph which is exemplary of the graphs common to metamaterials that deal with energy manipulation. This example provides the juxtaposition between specific charge q*, and frequency. For electromagnetic metamaterials for a certain frequency range, the material no longer is affected by the change in charge across the electric field. This region where the frequency becomes discontinuous is known as the band gap region.

A particularly notable example of MM functionality is cloaking, where MMs are designed to bend waves around them, or pass through an otherwise opaque object, effectively rendering the object invisible or undetectable by the stimuli [16, 36]. This cloaking ability is available in all four types of MMs. Such cloaking can be desired for optical applications to make the target visible for a camera or acoustic applications to make the target invisible to sonar. A combination of microstructure and macrostructure as well as selection of appropriate materials are effective in producing such effects.

As outlined in Table 1.1, MMs have a diverse range of functionalities which is underscored by their structures, properties, and applications. This serves to leave discussion open to delve deeper into the material categories, manufacturing methods, and limitations highlighting the importance of advanced modeling and manufacturing techniques to overcome the challenges of producing such complex geometries. As the new manufacturing methods are being developed at a rapid pace and many novel manufacturing methods are entering mainstream, the metamaterial

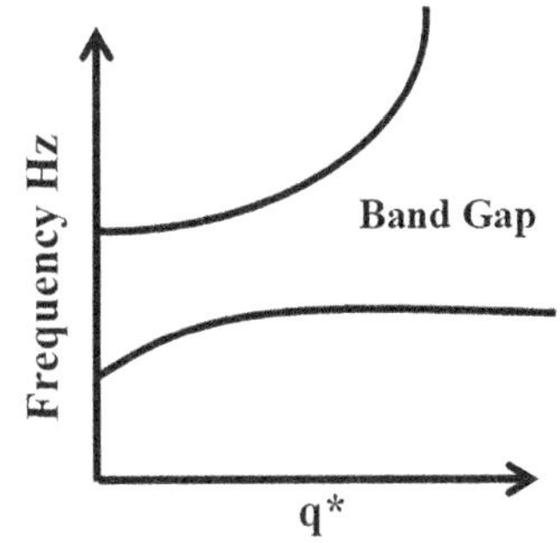

Fig. 1.4 Band gap graph depicting the separation of specific charge q*, and frequency in Hertz where the band gap formed is a result of a specific material property resulting in a specific discontinuity in the frequency

field is benefitting from the new capabilities of these methods to manufacture metamaterials. Highly damping soft materials are a focus of many applications. Metamaterials comprising these materials also show a rapid pace of innovation due to the ability to additively manufacturing them using several commercially available methods and the available mechanical properties over a wide range of loading conditions. The following chapters will discuss in depth the structure and properties of a wide range of metamaterials. The review will help in identifying the future trends in applications and research directions in these materials.

1.3 Summary

Metamaterials present a great promise for many applications where traditional materials are not able to demonstrate the appropriate properties. Their ability to offer unprecedented control over material properties through their micro- and macro-structure is a topic of interest in research and in application development. Through a detailed exploration of the core principles, classifications, and functionalities of MMs, a deeper examination of their specific applications, working mechanisms, and manufacturing methods can be conducted. The potential of metamaterials to revolutionize various industries is profound. In aerospace applications, these materials promise to enhance functionality leading to lighter, stronger, and more fuel-efficient aircraft. In the field of diagnostic medicine, the versatility of imaging techniques facilitated by MMs could significantly improve diagnostic accuracy and patient outcomes through clearer noninvasive imaging solutions. As the boundary is continuously pushed of what is possible with these materials, the transition from laboratory findings to mainstream technology is increasingly becoming reality. The evolution of MMs not only underscores their theoretical promise but also demonstrates tangible progress in applying these materials in real-world settings.

References

1. Tawalbeh, M., Khan, H. A., Hajar, A. A., & Al-Othman, A. (2022). Applications of metamaterials. In A.-G. Olabi (Ed.), *Encyclopedia of smart materials* (pp. 11–24). Elsevier.
2. Iannace, G., Ciaburro, G., & Trematerra, A. (2021). Metamaterials acoustic barrier. *Applied Acoustics, 181*, 108172.
3. Li, B., Liang, Y., Xu, X., Chen, J., & Lin, Y.-S. (2023). Tunable terahertz metamaterial using electric split-ring resonator with refractive index sensing characteristic. *Physica E: Low-dimensional Systems and Nanostructures, 151*, 115740.
4. Al-Othman, A., Tawalbeh, M., Temsah, O., Khan, H. A., & Hajar, A. A. (2022). Metamaterials: Classifications and characteristics. In A.-G. Olabi (Ed.), *Encyclopedia of smart materials* (pp. 46–56). Elsevier.
5. Wagner, M. A., Schwarz, F., Huber, N., Geistlich, L., Galinski, H., & Spolenak, R. (2022). Deformation-induced topological transitions in mechanical metamaterials and their application to tunable non-linear stiffening. *Materials & Design, 221*, 110918.

6. Pashaeiadl, H., Naserpour, M., & Zapata-Rodríguez, C. J. (2018). Scattering of electromagnetic waves by a graphene-coated thin cylinder of left-handed metamaterial. *Optik, 159*, 123–132.
7. Gao, J., Cao, X., Xiao, M., Yang, Z., Zhou, X., Li, Y., Gao, L., Yan, W., Rabczuk, T., & Mai, Y.-W. (2023). Rational designs of mechanical metamaterials: Formulations, architectures, tessellations and prospects. *Materials Science and Engineering: R: Reports, 156*, 100755.
8. Dong, H.-W., Zhao, S.-D., Wang, Y.-S., Cheng, L., & Zhang, C. (2020). Robust 2D/3D multipolar acoustic metamaterials with broadband double negativity. *Journal of the Mechanics and Physics of Solids, 137*, 103889.
9. Arjunan, A., Baroutaji, A., Robinson, J., & Wang, C. (2022). Characteristics of acoustic metamaterials. In A.-G. Olabi (Ed.), *Encyclopedia of smart materials* (pp. 35–45). Elsevier.
10. Sakovsky, M., & Ermanni, P. (2020). A thin-shell shape adaptable composite metamaterial. *Composite Structures, 246*, 112390.
11. Lepidi, M., & Bacigalupo, A. (2018). Multi-parametric sensitivity analysis of the band structure for tetrachiral acoustic metamaterials. *International Journal of Solids and Structures, 136-137*, 186–202.
12. Zhao, A., Liu, C., Zheng, Y., Zhang, A., He, P., Chen, H., Wu, G., Zhang, M., Wu, T., & Gu, G. (2023). Evaluation of shock migration performance for a multi-stable mechanical metamaterial. *Composite Structures, 321*, 117312.
13. Zhou, X., Ren, L., Song, Z., Li, G., Zhang, J., Li, B., Wu, Q., Li, W., Ren, L., & Liu, Q. (2023). Advances in 3D/4D printing of mechanical metamaterials: From manufacturing to applications. *Composites Part B: Engineering, 254*, 110585.
14. Wu, L., Xue, J., Tian, X., Liu, T., & Li, D. (2023). 3D-printed metamaterials with versatile functionalities. *Chinese Journal of Mechanical Engineering: Additive Manufacturing Frontiers, 2*(3), 100091.
15. Ullah, N., Shabiul Islam, M., Hoque, A., Yong, W. H., Soliman, M. S., & Islam, M. T. (2024). A compact-sized four-band metamaterial-based perfect absorber for electromagnetic energy harvesting applications. *Optics & Laser Technology, 168*, 109836.
16. Roh, Y., Lee, S.-H., Kwak, J., Song, H. S., Shin, S., Kim, Y. K., Wu, J. W., Ju, B.-K., Kang, B., & Seo, M. (2022). Terahertz imaging with metamaterials for biological applications. *Sensors and Actuators B: Chemical, 352*, 130993.
17. Patil, K. S., & Rufus, E. (2022). Design of bio-implantable antenna using metamaterial substrate. *Wireless Personal Communications, 124*(2), 1443–1455.
18. Zhong, H., Tian, Y., Gao, N., Lu, K., & Wu, J. (2021). Ultra-thin composite underwater honeycomb-type acoustic metamaterial with broadband sound insulation and high hydrostatic pressure resistance. *Composite Structures, 277*, 114603.
19. Askari, M., Hutchins, D. A., Watson, R. L., Astolfi, L., Nie, L., Freear, S., Thomas, P. J., Laureti, S., Ricci, M., Clark, M., & Clare, A. T. (2020). An ultrasonic metallic Fabry–Pérot metamaterial for use in water. *Additive Manufacturing, 35*, 101309.
20. Miao, Z.-H., & Wang, Y.-Z. (2022). Nonreciprocal transmission of non-collinear mixing wave in nonlinear elastic wave metamaterial. *Journal of Elasticity, 150*(2), 341–366.
21. Ma, H., Wang, K., Zhao, H., Shi, W., Xue, J., Zhou, Y., Li, Q., Wang, G., & Yan, B. (2022). Energy dissipation and shock isolation using novel metamaterials. *International Journal of Mechanical Sciences, 228*, 107464.
22. Su, Y., Xu, X., Shi, J., & Huang, G. (2021). A 3D mechanism-driven hexagonal metamaterial: Evaluation of auxetic behavior. *International Journal of Mechanical Sciences, 209*, 106699.
23. Bhuwal, A. S., Liu, T., Ashcroft, I., & Sun, W. (2021). Localization and coalescence of imperfect planar FCC truss lattice metamaterials under multiaxial loadings. *Mechanics of Materials, 160*, 103996.
24. Bai, J., Pang, Z., Shen, P., Chen, T., Shen, W., Wang, S., & Chang, S. (2021). A terahertz photothermoelectric detector based on metamaterial absorber. *Optics Communications, 497*, 127184.
25. Islam, M. S., Islam, M. T., & Almutairi, A. F. (2022). A portable non-invasive microwave based head imaging system using compact metamaterial loaded 3D unidirectional antenna for stroke detection. *Scientific Reports, 12*(1), 8895.

26. You, J.-H., & Park, K. (2021). Design and additive manufacturing of thermal metamaterial with high thermal resistance and cooling capability. *Additive Manufacturing, 41*, 101947.
27. Liu, J., Li, L., Xia, B., & Man, X. (2018). Fractal labyrinthine acoustic metamaterial in planar lattices. *International Journal of Solids and Structures, 132-133*, 20–30.
28. Lin, Y.-S., Yan, K., Yao, D., & Yu, Y. (2019). Investigation of electromagnetic response of terahertz metamaterial by using split-disk resonator. *Optics & Laser Technology, 111*, 509–514.
29. Sinha, P., & Mukhopadhyay, T. (2023). Programmable multi-physical mechanics of mechanical metamaterials. *Materials Science and Engineering: R: Reports, 155*, 100745.
30. Álvarez-Trejo, A., Cuan-Urquizo, E., Bhate, D., & Roman-Flores, A. (2023). Mechanical metamaterials with topologies based on curved elements: An overview of design, additive manufacturing and mechanical properties. *Materials & Design, 233*, 112190.
31. Poveda, R. L., & Gupta, N. (2016). Carbon nanofibers: Structure and fabrication. In *Carbon nanofiber reinforced polymer composites* (pp. 11–26). Springer.
32. Zeltmann, S. E., Poveda, R., & Gupta, N. (2016). *Environmental degradation of carbon nanofiber reinforced syntactic foams*. Springer International Publishing.
33. Poveda, R., Gupta, N., & Porfiri, M. (2010). Poisson's ratio of hollow particle filled composites. *Materials Letters, 64*(21), 2360–2362.
34. Lee, G., Lee, S.-J., Rho, J., & Kim, M. (2023). Acoustic and mechanical metamaterials for energy harvesting and self-powered sensing applications. *Materials Today Energy, 37*, 101387.
35. Jiao, P. (2023). Mechanical energy metamaterials in interstellar travel. *Progress in Materials Science, 137*, 101132.
36. Chen, Y., Mai, Y.-W., & Ye, L. (2023). Perspectives for multiphase mechanical metamaterials. *Materials Science and Engineering: R: Reports, 153*, 100725.

Chapter 2
Mechanical Properties of Metamaterials

List of Symbols

α	Angle of incidence
C_a	Speed of sound through a medium
E	Elastic modulus
Es, Ex	Directional elastic modulus
ε	Strain
εx, εy	X and Y directional strain constants
f	Frequency
$\vec{G}$	Periodic lattice vector
I	second moment of inertia
$\vec{k}$	Wave vector
k^*	Combined thermal conductivity behavior
k_g	Gas conductivity
k_s	Solid conductivity
$L_{x,y,z}$	Length in all directions
m_p	Mass per unit area
t	Thickness
$n_{x,y,z}$	Integers
$\vec{r}$	Position vector
$\overrightarrow{R_n}$	Lattice vector
S	Cross-sectional area
Θ	Angular geometric constraint
$w(x, t)$	Transverse beam deflection
ρ	Density
ρ_a	Density of acoustic medium
ρ^*	Apparent density
ρ_s	Solid density
$\sigma\infty$	Arbitrary stress
υ_{xy}	Poisson's ratio in x & y
V	Volume

N. Gupta, C. Beckwith, *Architected Metamaterials: Design Principles and Properties*, https://doi.org/10.1007/978-3-031-98330-6_2

ω Angular frequency
$\psi_{\vec{k}}$ Bloch amplitude

2.1 Introduction

Mechanical metamaterials (MMs) with the ability to manipulate physical properties have garnered significant attention in research due to their wide-ranging applications resulting from the interesting properties such as negative stiffness, negative dynamic mass density, and tailoring of damping ratio, among others [1–4]. Since these metamaterial microstructures are often designed as unit cells, which are repeated to fill space, analytical models and computer simulation tools such as finite element analysis (FEA) are widely used to estimate their mechanical properties. These materials take advantage of the deformation and displacement of the unit cell when interacting with a stimulus such as force, temperature, or acoustic wave to show the desired functionality.

The analytical approach of these materials uses the Bloch-Floquet theory to develop boundary conditions and find their dispersion curves [5]. This theorem provides a method to analyze the behavior of periodic systems through a dispersion relation which describes the reaction of a wave as it passes through a periodic crystalline structure. This is a classical mathematical model for mapping the behavior of a periodic structure. Bloch's theorem states that a periodic lattice needs to satisfy the relation [5]

$$\psi_{\vec{k}}\left(\vec{r}+\overrightarrow{R_n}\right)=e^{i\vec{k}\cdot\overrightarrow{R_n}}\psi_{\vec{k}}\left(\vec{r}\right) \tag{2.1}$$

The dispersion relation can then be established by the wave vector along the edges of the Brillouin zone [6–10]. Where $\psi_{\vec{k}}$ is the Bloch amplitude of a wave as it passes through the position vector and lattice field vector $\vec{r}$, and $\overrightarrow{R_n}$, respectively, within the bounds of $\vec{k}$ the vector of the incident wave. These resulting boundary conditions allow for the creation of the individual unit cells, which can be assembled to produce the material with desired properties. An example of this phenomenon would be the manipulation of a material's stiffness by introducing a chiral unit cell to create a specific acoustic bandgap [11, 12]. Figure 2.1 depicts (a) the simplified unidirectional periodic unit cell typically found in analytical solutions for elastic MMs [13], (b) the first four mode shapes for an elastic metamaterial starting from rest [6], and (c) the deformation and strain profile of a periodic bi-directional elastic MM [14]. While in literature most MMs are developed as periodic lattice structures, recent development in machine learning methods and advanced manufacturing techniques such as additive manufacturing (AM) has made it possible to study aperiodic structures, which are likely to be the focus in this field in the coming years.

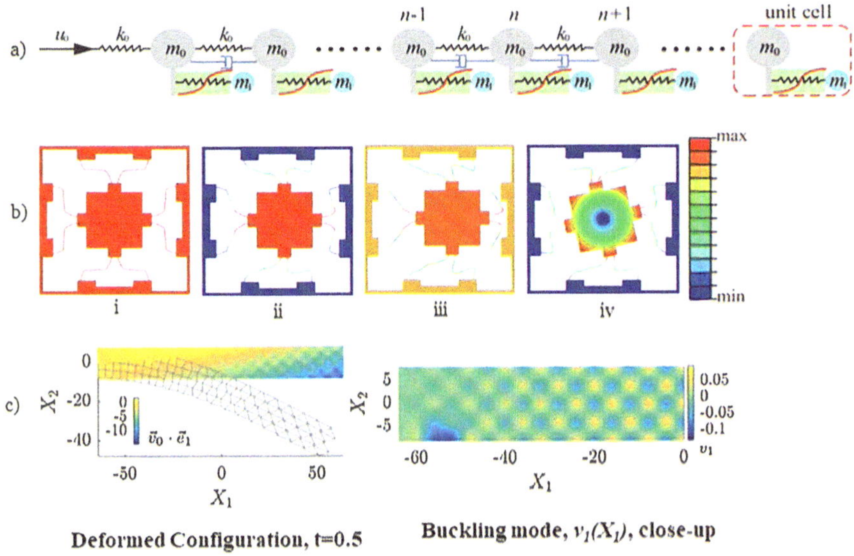

Fig. 2.1 (**a**) Simplified dynamic model of elastic MM for arbitrary periodic boundary n where k is a spring constant and m is a mass for provided initial displacement u_0 adapted from [13]. (**b**) Mode shape simulations of single unit cell of elastic MM for static (i) first translation (ii) second translation (iii) and torsion (iv) adapted from [6]. (**c**) Deformation simulation first mode shape for elastic MM across distances X_1 and X_2 adapted from [14]

Acoustic metamaterials have contributed to the development of elastic MMs. The occurrence of nonlocal interactions at various points over the lattice tends to have large effects on sound attenuation. This effect creates unexpected bandgaps which can be manipulated for specific applications [15–18]. Underwater applications are often the focus of elastic acoustic MMs. For example, metamaterials have been developed for actively adjusting sound pressure level through locally creating negative mass density [19–21].

The versatility of tunable properties in metamaterials are often the aspect which makes them such a widely sought replacement for traditional materials. For elastic MMs, a substantial portion of research deals with active control over a material's properties to meet the needs of niche applications. A popular method to accomplish this objective is to design a functional gradient of unit cells to target specific frequencies to create a bandgap [22], or through the use of a base composition whose rheology can be manipulated through some stimulant [23]. A novel approach to tuning material properties for elastic mechanical MMs used heat to actively tune the stiffness of their material such that they could manipulate incident flexural waves [24].

Band gap formation in electromagnetic MMs has been analytically examined using the plane wave expansion method to describe the motion of a three-dimensional periodic lattice [25].

For a three-dimensional medium, lattice basis vectors can be used to describe the structure periodicity. The plane wave expansion method used in this work is based on a method to solve Maxwell's fundamental equations of electromagnetism by ordering them into an eigenvalue problem. In this way each tensor and vector can be broken down into material properties and wave interactions to model the material's response. The plane wave expansion method describes the structure of a periodic material through the formation of a periodic lattice with the lattice vector [25].

$$\vec{G} = \frac{2\pi}{L_x} n_x \hat{i} + \frac{2\pi}{L_y} n_y \hat{j} + \frac{2\pi}{L_z} n_z \hat{k} \tag{2.2}$$

This lattice vector $\boldsymbol{G}$ represents the periodic structure that has material properties assigned to it through expansion using a Fourier series, while $\boldsymbol{L_x}$, $\boldsymbol{L_y}$, and $\boldsymbol{L_z}$ are the physical dimensions of the lattice for coordinates $\boldsymbol{n_x}$, $\boldsymbol{n_y}$, and $\boldsymbol{n_z}$. Once the material's spatial orientation is modeled through Eq. 2.2, each property can be calculated by integrating over the material's volume domain [25].

$$p(\vec{G}) = \frac{1}{V} \int_V p(\vec{r}) e^{-i\vec{G}\cdot\vec{r}} \, dV \tag{2.3}$$

From here it is possible to calculate the properties of each component $\boldsymbol{p(G)}$ in a periodic lattice structure by discretizing each component as a coordinate along $\boldsymbol{G}$ and summing them over the domain such as what is done in a Bloch solution for boundary conditions represented as the volume $\boldsymbol{V}$ across Bloch's position vector $\vec{\boldsymbol{r}}$. This work made a good showing for evaluating the elastic properties of three-dimensional periodic lattice structures through analytical models showing high convergence [25].

Understanding nonlinearities in elastic behavior is the other major focus area for elastic metamaterials. Developing models for nonlinear behavior and dissipation are paramount to understanding, controlling, and manipulating elastic waves [26]. Numerical analysis of the microarchitecture of these materials often simplifies these materials to damped mass nonlinear-spring system often arranged for single degree of freedom analysis under shock response analysis [27]. These studies often consider stretching and bending behaviors of materials and dispersion properties in the first Brillouin zone [28].

2.2 Auxetic Materials

Auxetic materials are a specific branch of mechanical MMs which possess a negative Poisson's ratio. The lattice geometries used in their structural design drives their mechanical properties in addition to the composition of the material [29–32]. Auxetic materials can be designed with many different lattice geometries, each of which produces a different result due to the differing effective stiffness [33–35].

Their microstructure design can result in interesting acoustic properties for vibration and sound transmission leading to the creation of acoustic lenses [36, 37]. Hence, auxetic MMs are primarily used in structural and dynamic applications such as robotic manipulators, lightweighting aircraft, and dampening impacts from automobiles [38–40]. Recent studies have manufactured these materials through additive manufacturing methods such as fused filament fabrication [41, 42]. The literature suggests these materials are primarily composed of 2D lattice structures and rarely venture into 3D structures, primarily due to the complexity in manufacturing 3D architectured materials by traditional methods.

There are four major areas of research surrounding these materials: (i) structures with tunable properties, (ii) bi-material auxetic structures for negative Poisson's ratio and negative coefficient of thermal expansion, (iii) buckling behavior of auxetic structures, and (iv) origami-inspired structures and their buckling behavior. Figure 2.2 provides examples of how auxetic materials appear and behave. Figure 2.2a presents representation of unit cells and periodically corresponding MMs composed of vertex-connected three-, four-, and six-sided rotational symmetry bi-material cells, and strut-connected cells of the same rotational symmetries [43]. Figure 2.2b presents an Ashby diagram depicting the specific energy absorption (SEA) with respect to density of many materials including auxetic metamaterials represented by the star [44]. As per the general description of the concept of Ashby diagrams, the general trend of the existing materials is from the bottom left to top right corner of the diagram. In comparison, the star represents high SEA at low density, which implies that these metamaterials can provide significant weight saving in energy absorption applications compared to the traditional applications. Studies have proposed several different metamaterial designs, which can be used to tune the materials for the requirements of an application.

A novel design of an auxetic origami metamaterial used a folded lattice structure [45]. Each layer of the structure was pressed from 0.3 mm thick sheets of brass using an additively manufactured die and adhered to one another using an adhesive. These layers change in angle creating a functionally graded lattice structure. Due to the oblique angle at which the material is loaded, the phenomena of stiffness recovery is observable during compression testing. Figure 2.3 illustrates the finite element analysis simulation and experimental results of this material's compressive capabilities. Figure 2.3a, c correspond to the first experiment ran with these materials where there was no change in angle between the layers, and Fig. 2.3b, d correspond to the experiment where the angle of each layer was changed. From the results, it can be observed that the functional grading of the lattice changes the stiffness and compressive strength of the material allowing the stiffness to increase as each layer is flattened under the compressive force.

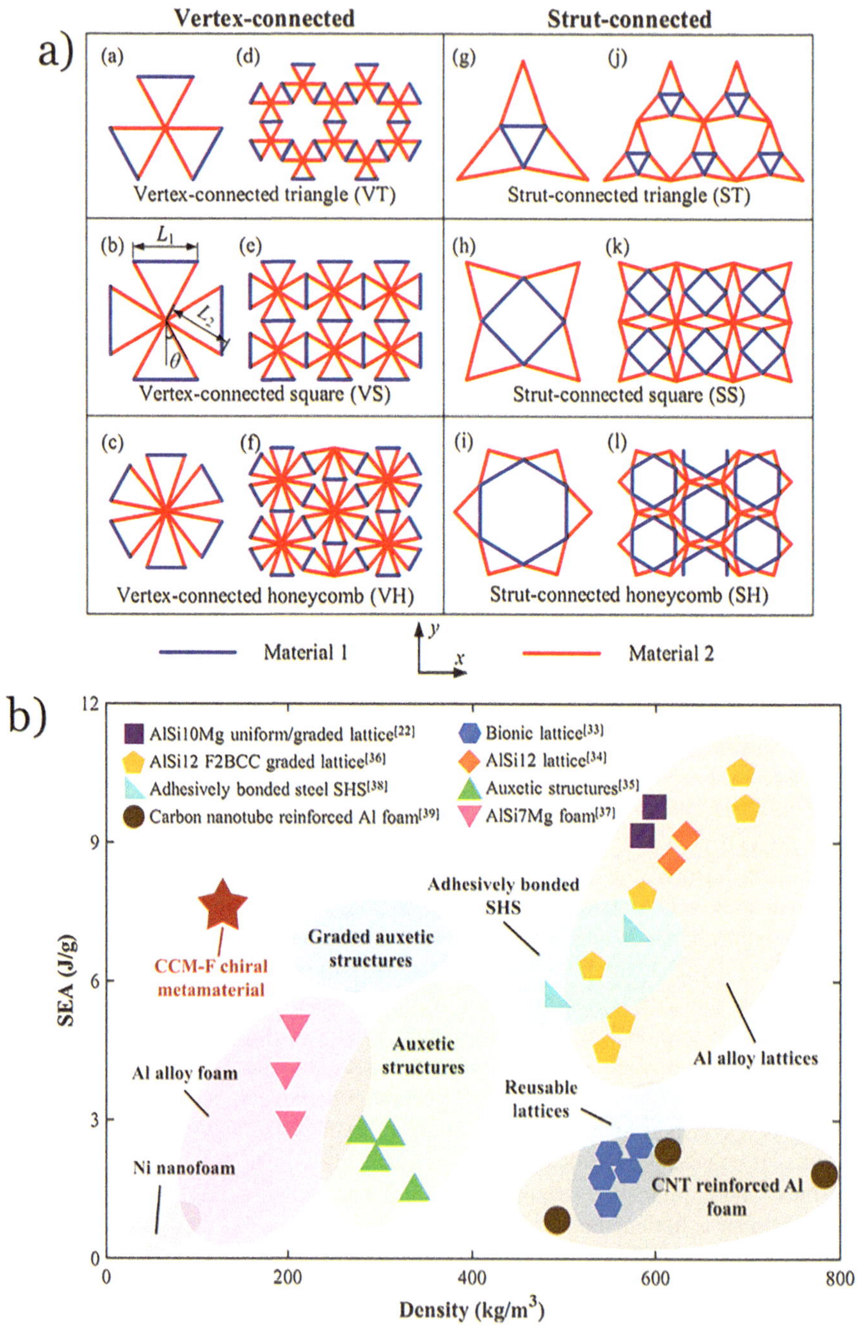

Fig. 2.2 (**a**) Bi-material auxetic metamaterial formation adapted from [43]. (**b**) Auxetic metamaterial Ashby diagram for specific energy absorption against density adapted from [44]

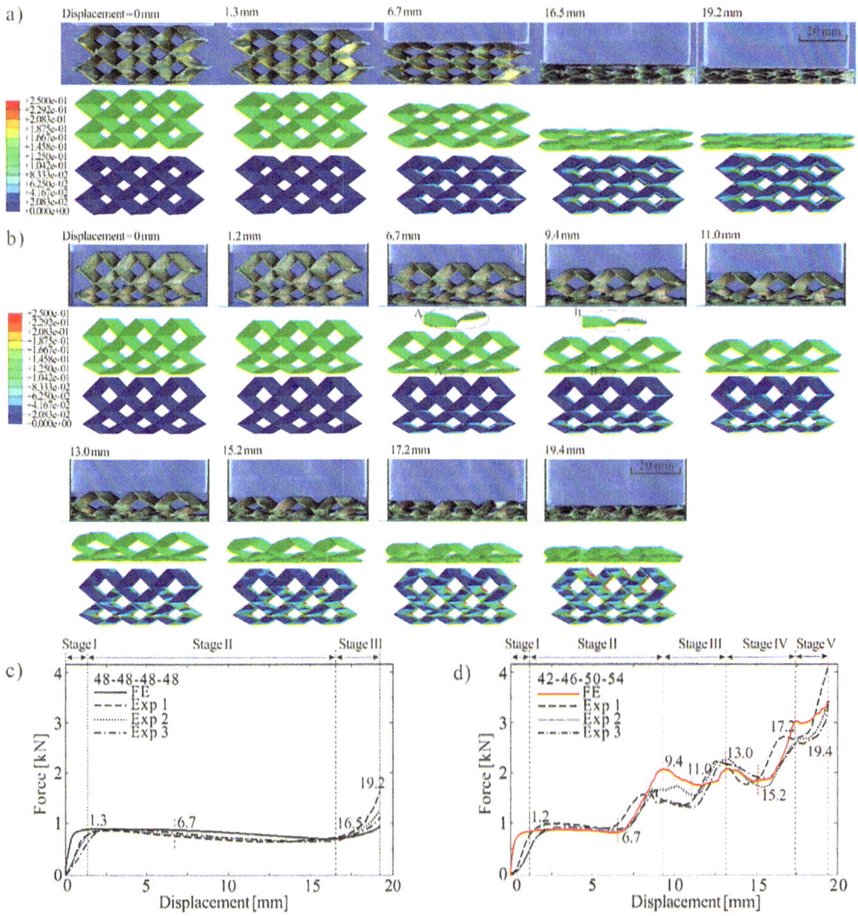

Fig. 2.3 (**a**) FEA and experimental results from compression test with uniform lattice layers. (**b**) FEA and experimental results from compression test with gradient structure applied to the layers of the lattice. (**c**) Graphical results from FEA and compression testing of uniform lattice structure. (**d**) Graphical results from FEA and tensile testing of gradient lattice structure adapted from [45]

2.3 Tunable Properties

The structural design parameters for auxetic materials include unit cell geometry, arrangement, and material composition to achieve the desired negative Poisson's ratio as well as elastic modulus. Understanding the effects of chirality and angular position of a unit cell can help in designing novel structural response. For example, a material with a typically positive Poisson's ratio can be used to design a structure that demonstrates a negative Poisson's ratio [46–48]. MMs whose properties are chosen in the design phase typically do not have the ability to alter those properties after manufacturing. These MMs retain the ability to change their material

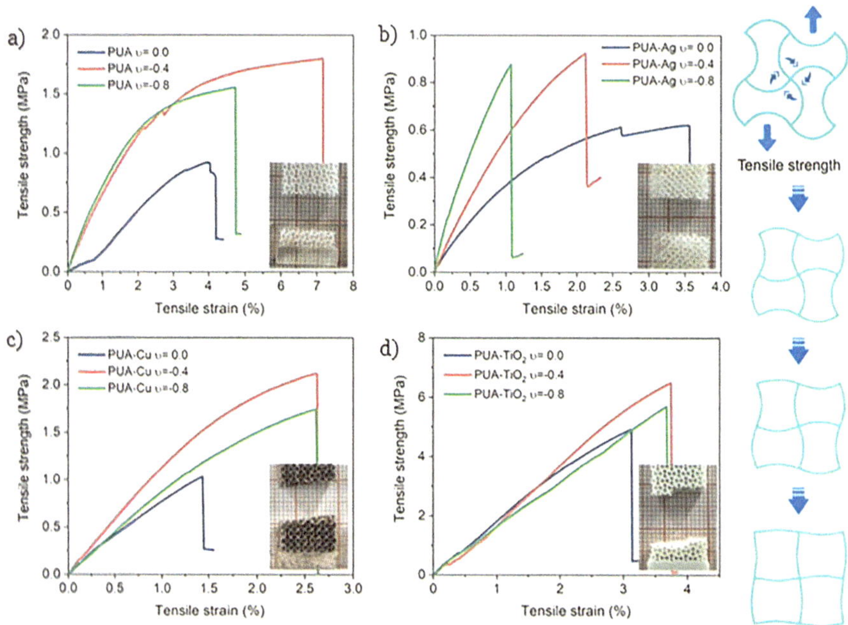

Fig. 2.4 Tensile stress-strain curves for negative Poisson's ratio metamaterial. Four tests composed of polyurethane acrylate both as pure polymer (**a**) and with various additives including gold (**b**), copper (**c**), and titanium oxide (**d**). The right depicts the transformation mechanism from the negative Poisson's ratio effect adapted from [60]

properties in response to various stimuli [49, 50]. For example, Poisson's ratio sine-switching is a popular topic in literature wherein when presented with significant compressive stress the sign of Poisson's ratio will switch between positive and negative. This phenomenon occurs after a certain threshold of pressure is reached and the material becomes fully dense [51–53]. Figure 2.4 provides further insight into the mechanism behind negative Poisson's ratio materials by providing four examples of lattices stress-strain curves. These lattices were constructed with negative Poisson's ratio formations and were composed of polyurethane acrylate with metal additives such as gold, copper, and titanium oxide. Figure 2.4 provides the transformation mechanism behind negative Poisson's ratio effect causing the unit cell to unfurl and expand as tension is applied to it. In this instance it can be observed that the metal additives hampered the performance of the polymer lattice having half the maximum stress as gold is added [60].

A tunable auxetic MM was examined and validated through additive manufacturing by optimizing the topology of the unit cells based on the Poisson's ratio and elastic modulus [54]. The second-order hierarchical structure presented in this work combined analysis of two- and three-dimensional lattice structures such that the resulting topology possessed positive elastic properties.

Poisson's ratio is a fundamental material property which describes the behavior of a material as it deforms in response to applied strain. Most materials exhibit a positive Poisson's ratio, meaning they reduce its diameter as they are pulled in tension. Metamaterials often intentionally or unintentionally exhibit a negative Poisson's ratio, meaning they will expand in tension and contract in compression. The expression for the macroscopic Poisson's ratio of this quadrilateral honeycomb structure is given as [54]:

$$\upsilon_{xy} = -\frac{\varepsilon_x}{\varepsilon_y} = \frac{1}{\tan\theta_1 \tan\theta_2} \tag{2.4}$$

This equation makes some assumptions such as assuming the struts of the lattice to be Euler beams and that the displacement and rotation angle of the unit cell can be calculated. Where υ_{xy} is Poisson's ratio, ε_x and ε_y are the transverse and longitudinal strains, and θ_1, and θ_2 are the angles of lattice struts closest to the parallel of the force. By using this equation, the geometric conditions can be related back to the directional elastic modulus [54].

$$\frac{E_y}{E_s} = \frac{\sigma_\infty}{\varepsilon_y} = \frac{\left(\tan^2\theta_1 \cos\theta_2 + \tan^2\theta_2 \cos\theta_1\right)}{\left(\tan\theta_1 + \tan\theta_2\right)}\left(\frac{t}{l}\right)^3 \tag{2.5}$$

This relation uses the same angles as before, relating to the geometric conditions of the unit cell but adds on t and l relating to the thickness of the strut and arclength of the unit cell. Herein the ratio of elastic modulus in the y direction is solved but the modulus in the x direction takes the same approach with an extra term of $\tan^2\theta_1\tan^2\theta_2$ in the denominator for an arbitrary stress and strain along the same angle $\frac{\sigma_\infty}{\varepsilon_y}$ yield that the deformation can be described entirely in geometric terms. Ultimately following this theoretical framework, lattice structures were designed and additively manufactured. These structures were validated by compression testing, confirming the expected metamaterial properties based on the design of the material.

2.4 Postbuckling Behavior

Extensive research is ongoing into understanding the buckling behavior of MMs. Compression testing of these materials reveal interesting responses regarding the behavior of the stiffness of these materials. An example of postbuckling behavior can be seen in Fig. 2.5, adapted from [55]. A shape memory polymer induces a recovery in the material's stiffness in response to various levels of stress and applied temperature [55–57]. It can also be observed from this figure that thermal stress over a range of temperatures produces the same change in stiffness over the length

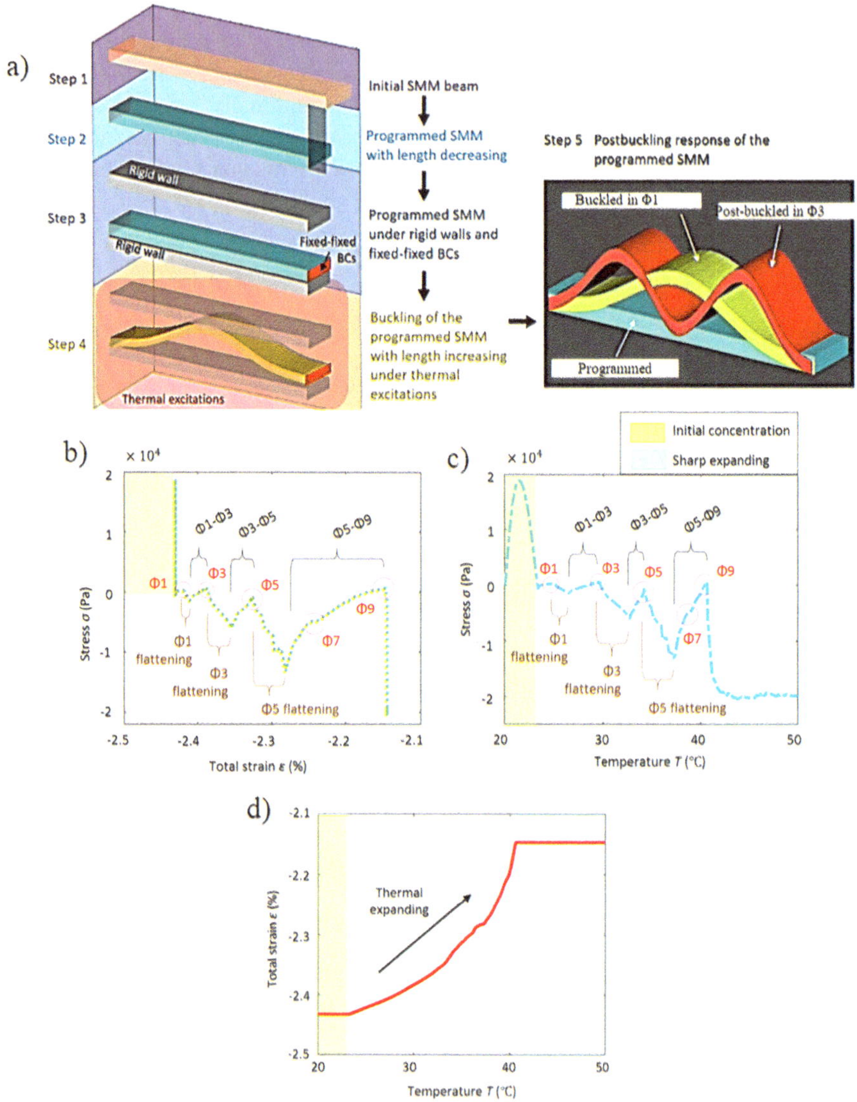

Fig. 2.5 (**a**) Theoretical design of a shape memory metamaterial postbuckling behavior with changes in stiffness. (**b**) The stress on the lattice under compression with respect to the total strain as a percent. (**c**) The stress on the lattice under compression with respect to the temperature. (**d**) The total strain experienced by the lattice as a function of the temperature adapted from [55]

of the material. The formation of the peaks in the stress-temperature diagrams indicates the lattice structure progressively collapsing across multiple layers and recovering stiffness as the structure densifies. The unit cell strut thickness, periodicity, and chirality are among the parameters studied for their role on the compressive deformation of the MM [44, 58–60]. The literature suggests the local plasticity resulting from the individual unit cells, which strengthens the material at periodic boundaries of the cells, as an explanation for this variable stiffness [61, 62].

Recently, two novel approaches have been taken to study the specific energy absorption of these material [63, 64].

Acoustic waves result in mechanical deformation of materials. The understanding of the amount of sound dampening the material provides is a direct relation to the mechanical performance and energy the material can absorb by relating this back to the elastic modulus of the material. In this instance once the frequency range has been breached, the material will have been excited to the motion state that resembles its postbuckling state. The first approach examined the sound transmission loss performance of materials, finding the material which defines the sound transmission loss (STL) of the material mass and density of the lattice. STL is used as a metric for describing the acoustic performance of a material; herein the STL was modeled for plane waves traveling through an infinitely long and thin panel [63].

$$STL(\omega) = 10\lg\left(1 + \left(\frac{m_p \omega \cos\alpha}{2\rho_a c_a}\right)^2\right) \approx 20\lg\left(m_p f\right) - 42 \tag{2.6}$$

The optimal STL range found through this analysis ranged between 50 and 150 Hz. The second analysis approach examined the auxetic lattice for its mechanical properties through compression testing. The energy absorbed is measured through the compressive strain over the length of the lattice. Finding thickness of the lattice struts correlates with the energy absorbed by the structure, an important conclusion to draw for these materials as the largest category of MMs surveyed pertains to energy absorption.

2.5 Origami-Inspired Auxetic Metamaterials

Origami-inspired auxetic MMs are a variety of mechanical MMs which make use of their folding shape to alter their stiffness and create the buckling behaviors previously discussed. Taking their lattice design and inspiration from the Japanese technique for folding paper into complex shapes, these auxetic materials are developed using their buckling behavior to fold the lattice on top of itself. These materials are often validated after manufacturing from layers like layers of thin sheets of paper [65–67]. Adjusting the thickness between the layers allows for a gradient stiffness to be created among the successive layers of the material. This allows multiple peeks to appear in the stress-strain curve when compressing the material [68–70].

Across these materials, the folding creates a relationship between Poisson's ratio, elastic modulus, and density. This relationship is created by the self-locking connections formed when the lattice is subject to large deformations [71, 72]. These

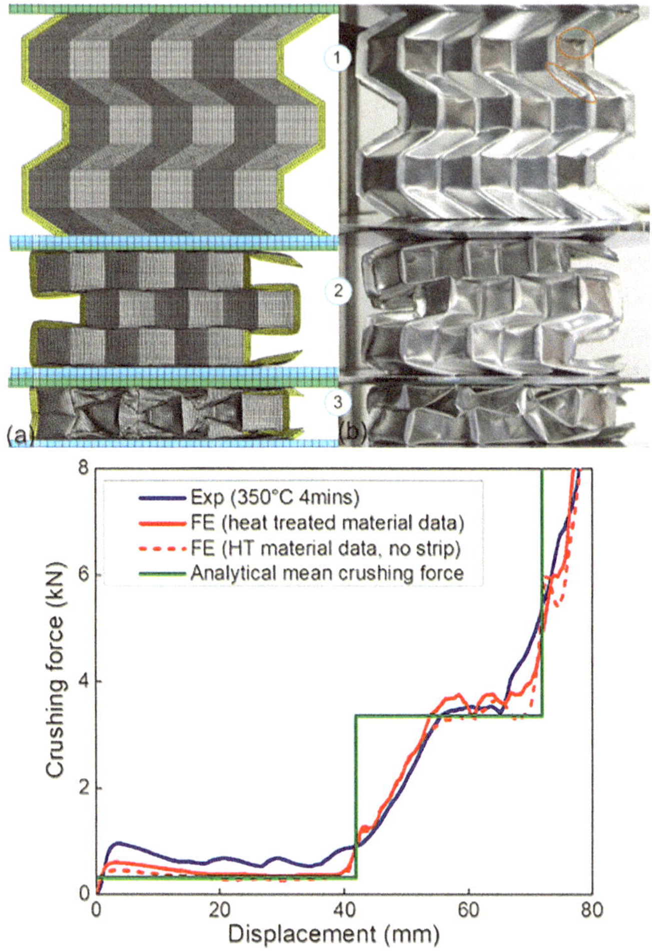

Fig. 2.6 Crushing behavior of origami auxetic metamaterial with force displacement chart adapted from [68]

metamaterials provide an opening into further understanding the properties that are relevant to engineering applications such as expanding structures, deployable structures, and folding structures [45]. Figure 2.6 provides an example of the buckling behavior of origami-inspired auxetic MMs. It can be observed from here that the same stair stepping effect occurs as in the buckling behavior due to the portions of the lattice folding on each successive layer rather than fully failing providing a second stage for the MM's life span [68].

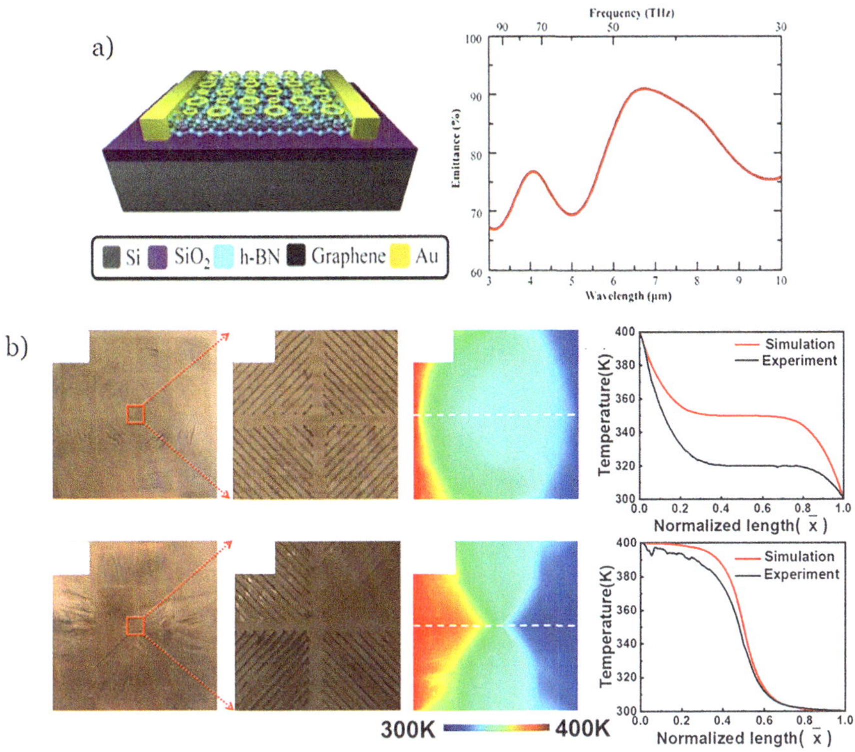

Fig. 2.7 (**a**) Thermal emission metamaterial adapted from [78]. (**b**) Heat flux control material adapted from [80]

2.6 Thermally Programmable Metamaterials

Thermal metamaterials are developed specifically to deal with heat flux management pertaining to convective, conductive, and radiative heat sources [73, 74] for thermal emission, expansion, and flux control. Many thermal MMs are composed of ceramics, composites, and metals, which are either 3D printed or machined [75]. Thermal expansion, conductivity, and emissivity under various temperature conditions as well as changes in elastic behavior are the primary properties evaluated in these materials [76, 77].

Figure 2.7 provides a graphical representation of the three main categories of thermal metamaterials. Figure 2.7a presents a thermal emission metamaterial whose emissivity was determined by the size and placement of graphene filament rings distributed throughout the surface of the emitter [78]. Figure 2.7b depicts a heat flux controlling metamaterial where the engraved pattern acts as a siphon or a shield directing the flow of heat either toward or away from the center of the material showing the effectiveness for metamaterials to act in a multifunctional sense to either shield from or absorb heat [79]. Figure 2.8 depicts an Ashby graph plotting

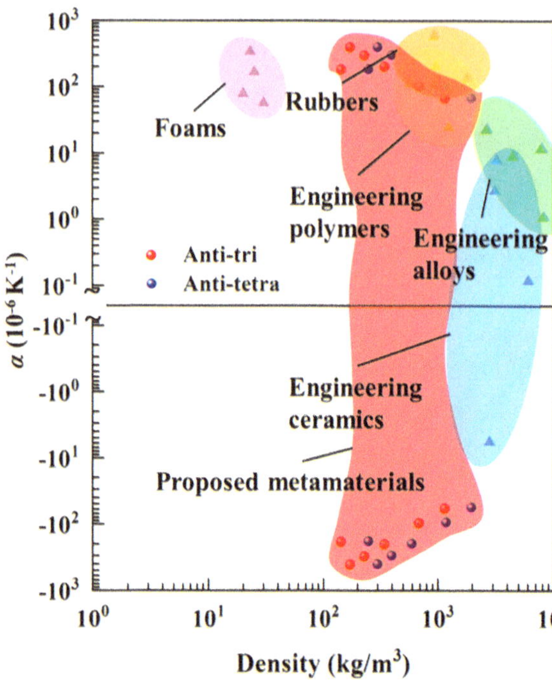

Fig. 2.8 Thermal expansion Ashby graph for multiple materials with the largest region representing the distribution of metamaterials with manipulatable thermal expansion coefficient for a range of densities adapted from [79]

the density of various materials against their coefficient of thermal expansion (CTE) compared to metamaterials, whose densities may remain similar to that of polymers. Most of the engineering materials fall in a narrow density range. Foams are notably lower density compared to the solid materials, but their CTE is very high. Despite metamaterials being composed of invar and steel, but whose CTE take up a wider range than any other common materials, their CTE goes deep into the negative range [80]. It is noted that most applications require the ability to tailor the properties over a wide range rather than having the lowest or the highest properties. The tailoring of the properties to the precise level predicted by the simulations provides the highest quality of metamaterial for a particular application.

2.7 Bi-Material Auxetic Metamaterials

Bi-material auxetic MMs are tunable auxetic MMs that typically blur the line between thermal MM and mechanical MM. They are typically fabricated from two materials with differing CTEs that are arranged in an auxetic lattice to create not

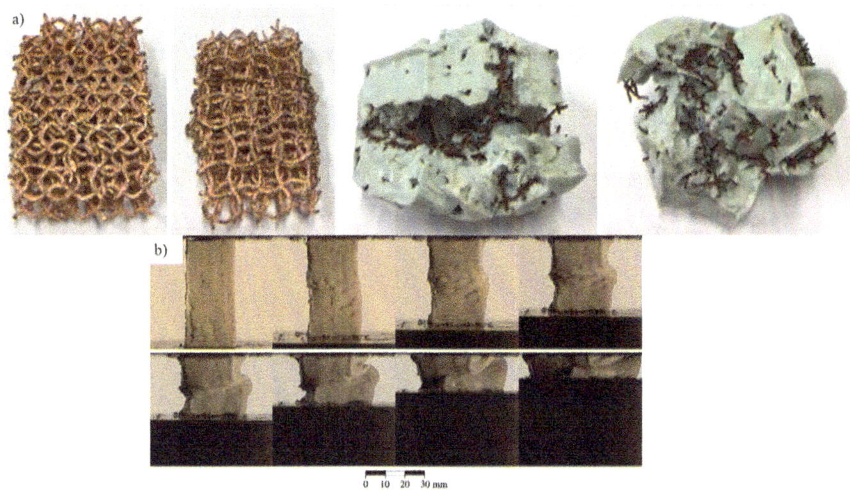

Fig. 2.9 Bi-material auxetic structure featuring (**a**) deformed results of graded copper additive manufacturing lattice structure, and copper silicone structure (**b**) deformation process of copper silicone structure adapted from [89]

only a negative Poisson's ratio but also a negative CTE [81–83]. Two different metals or polymers may be combined, where the outer struts of the unit cells may be composed of one material and the inner rings may be composed of the other material. The relationship between Poisson's ratio, stiffness, CTE and the relative density of the material has been studied primarily for 2D lattices, but some examples of 3D structures have also been found studied [84–86].

Often validated through compression testing of an additively manufactured sample these materials show stable results comparable to most widely available engineering materials but provide additional utility in their designed functionality [87–89]. Li et al. [90] analyzed on auxetic MMs in compression for both negative Poisson's ratio and negative thermal expansion effects. The material was fabricated with vat polymerization additive manufacturing method. Building upon the previous work [91], they built a 2D hierarchical star-square structure to conserve the stiffness and CTE of their lattice structure. The same is true for lattices using a strut-connected configuration. Figure 2.9 provides an example of a novel bi-material auxetic MM which used a powder bed fusion manufactured copper as a reinforcement lattice around silicon. The copper lattices were produced with uniform distribution and graded density. The combinations of the copper lattices with and without silicone were assessed for their compressive strength and it was demonstrated that the silicone-copper combination with the graded structure is more than either of the two materials on their own [89]. This material shows bi-material auxetic materials can be manufactured in multiple varieties and such that the combination of materials offers increased mechanical performance more than the two materials individually.

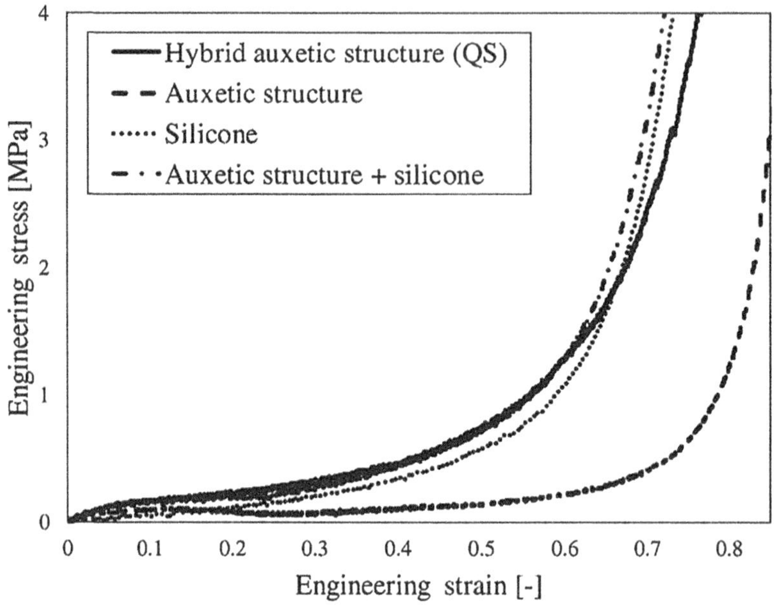

Fig. 2.10 Stress-strain graph of metamaterial compression results with copper lattice, silicone block, and copper lattice inside of silicone adapted from [89]

Figure 2.10 shows the resulting stress-strain graph of the MM; the hybrid material here has the highest slope whereas the pure materials show reduced performance.

2.8 Thermal Expansion

In nature it can be observed that as a material heats it both expands and becomes less stiff. Thermal expansion MMs are closely related to auxetic MMs using two conflicting materials and specific geometries to create a controllable thermal expansion coefficient CTE. This means that the resulting thermal MMs, depending on how they are manufactured, have a degree of control over their expansion and stiffness due to heat [43, 92]. There are multiple geometric formations which can be used to create this effect, all of which use a patterned lattice of materials strategically placed within the lattice to control the thermal expansion coefficient and evaluate the stiffness of the material [80, 93, 94]. As these materials are made of thin lattices, they are often most suited for light weight applications such as thermal insulator and thermal stress regulation [95, 96]. Figure 2.11 adapted from [80] provides an example of the experimental testing setup necessary to accurately measure the material properties relevant to this material. The thermal strain along a gradient of temperatures will show the usefulness of these lattices in determining the optimal loading conditions of the thermal MM. The tensile properties will provide optimal loading

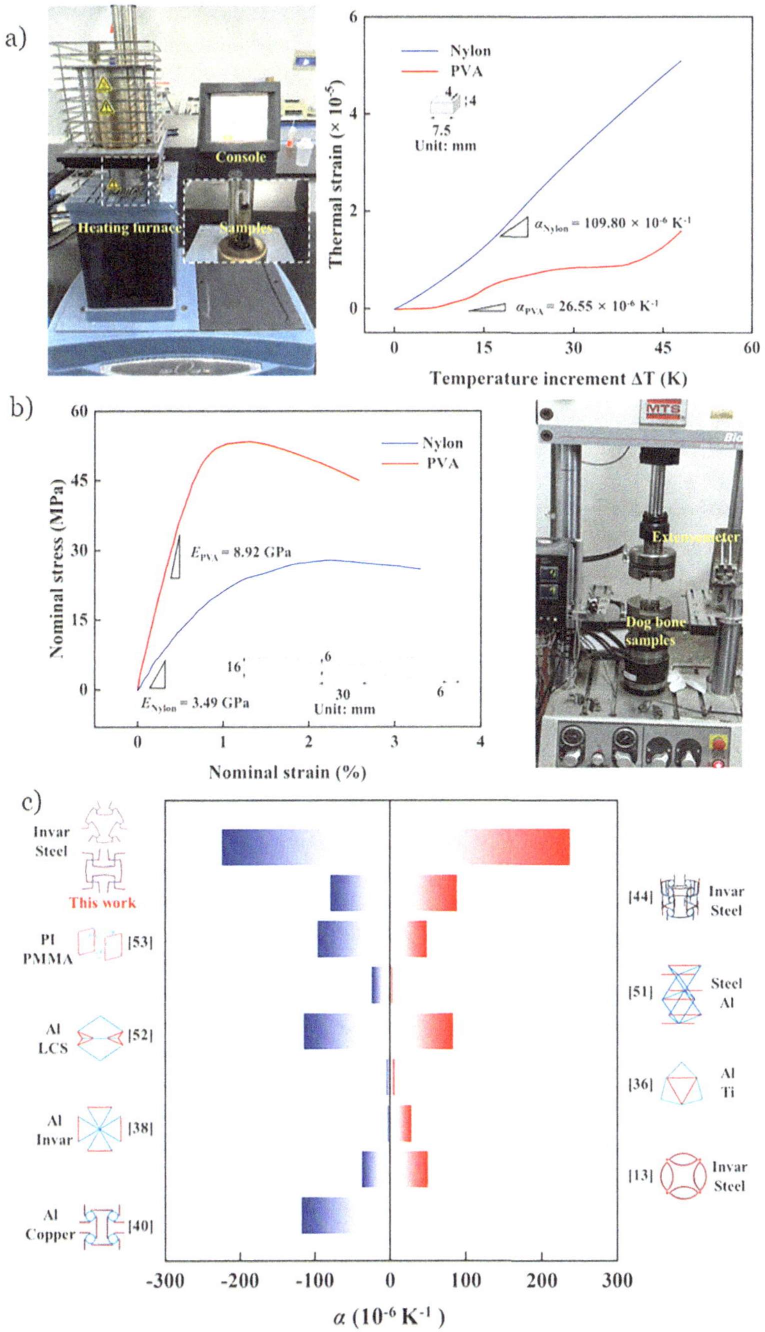

Fig. 2.11 (**a**) Experimental set up and test of nylon and PVA lattice structures for thermal strain. (**b**) Experimental set up and results for nylon and PVA lattices in strain tension. (**c**) Thermal expansion comparison of multiple lattice structures and material combinations adapted from [79]

conditions for the lattice. Figure 2.11c provides examples of various unit cells and material combinations for thermal expansion thermal MMs where the specific arrangements in the unit cells can be shown to produce positive or negative CTE despite the same composition being used across material, particularly in the case of steel-invar combinations where three results are shown each with differing expansion results.

A unique thermal expansion thermal MM-based approach was developed based on the Hoberman mechanism [97]. This thermal MM uses a combination of two materials such that one has a higher CTE than the other. The two different rates at which these materials expand due to heat produce an interesting effect in the material resulting in an uneven thermal expansion across a fully patterned lattice leading to sections contracting and others expanding in finite element simulation under the same applied stimulus. This material was simulated through five material combinations, each of which follows a similar trend demonstrating the effectiveness of this geometric combination through multiple material types such as plastic and metals. Figure 2.12 depicts the thermal expansion simulation and the stress-strain simulations of the theorized Hoberman MM illustrating the material arrangement and the produced negative thermal expansion.

2.9 Heat Flux Control

The manipulation of heat, driving it through a material and guiding it to a destination, is the goal of heat flux manipulation thermal MMs. In principle these materials expand upon Fourier's law of heat conduction by considering levels of hierarchies, transport phenomena, and variational principals which would alter the flow of heat within a material [98]. Many of these materials are composed of ceramics, composites, and metals such as silica and tungsten-based ceramics, alumina microtubes, and copper alloys [99, 100]. These materials are primarily developed as thermal insulators and isolators to drive heat away from sensitive components [79, 101]. Figure 2.13 offers an example of these materials through two 4D printed shape memory lattice structures evaluated for their conductance at multiple strain levels [101]. These lattices are then assessed under compression where they are heated to observe their shape memory capabilities to reform after mechanical deformation. The octet-truss and Kelvin cell depicted in Fig. 2.13 showed effective conductance improvements from 3.23 to 10.48 mW/K, and 4.41 to 14.52 mW/K at strain rate from 0% to 70%, respectively.

A study of the cooling capacity of a heat flux thermal MM manufactured through powder bed fusion additive manufacturing process using 316 stainless steel as the feed stock [102].

The thermal conductivity of the metamaterial is a fundamental property for observation. As heat is absorbed and dissipated by the material, this can be tracked through several volumetric methods. Fourier's law of heat conduction is a classical method to describe the movement of heat across a gradient. This law describes the movement of heat through a homogenous solid; however, in metamaterials this is expanded upon

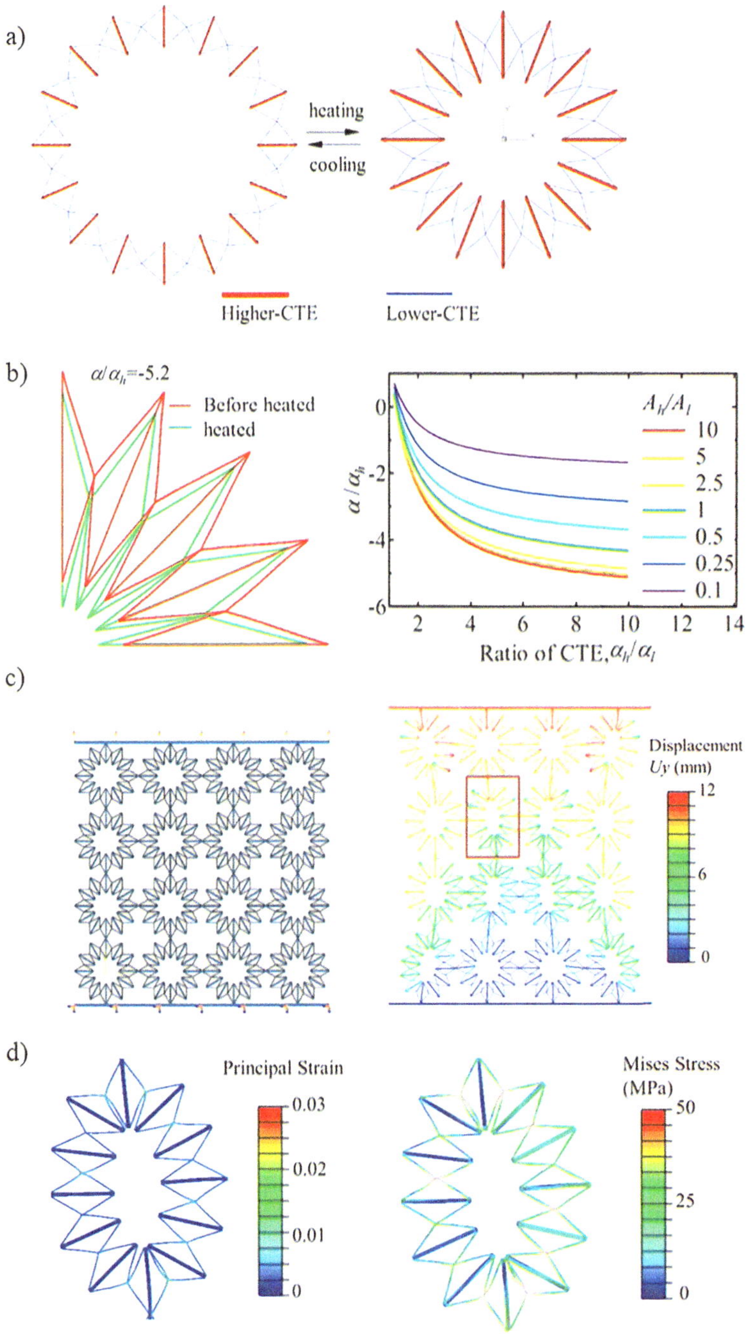

Fig. 2.12 (**a**) Theorized bi-material Hoberman metamaterial. (**b**) Simulated results of thermal expansion on Hoberman metamaterial. (**c**) Stress-strain testing results from simulated MM. (**d**) The principal stress and strain distributions of the individual unit cells adapted from [97]

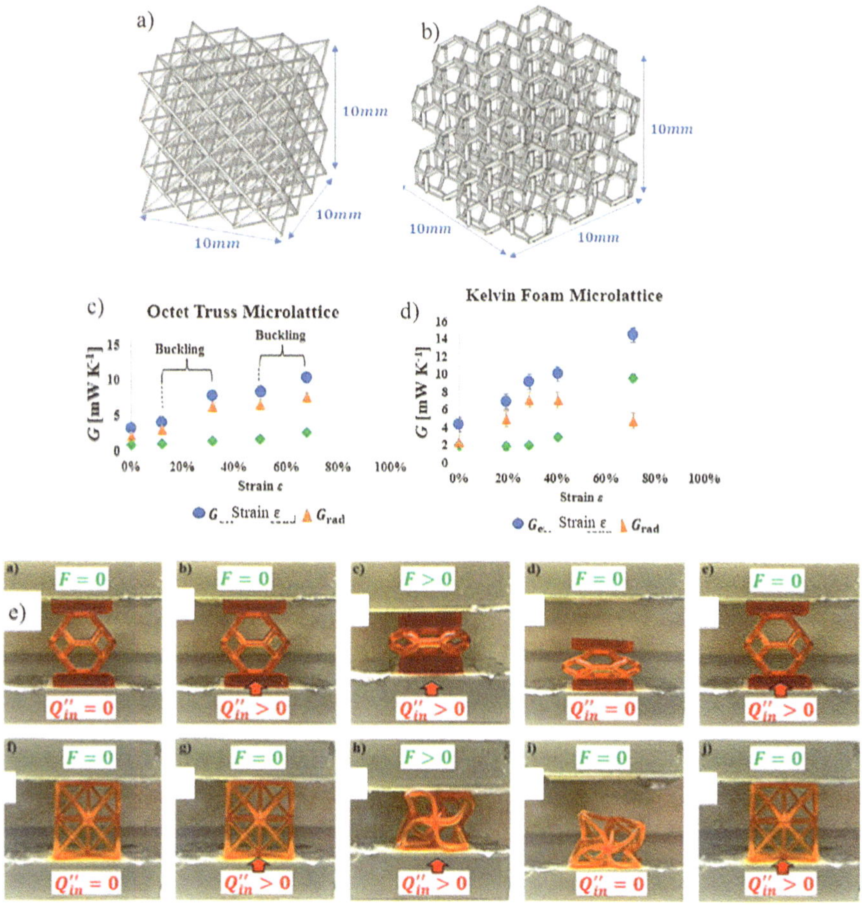

Fig. 2.13 4D metamaterial printed for shape memory retention through application of heat after compressive deformation adapted from [98]. (**a**) and (**b**) Depict the single unit cells and lattice structures used, composed of octet truss and kelvin foam cells, respectively. (**c**) and (**d**) Depict the unit cell and lattice performance under differing strain percents showing the correlation between increased thermal conductivity at higher strains. (**e**) Demonstrates the effectiveness of the material's shape memory as temperature changes past the glass transition temperature of the plastic the material returns to its initial shape after deformation [101]

often similarly to that of composite materials and open cellular materials [103–105]. This thermal MM was fabricated as a cellular lattice material to have the thermal conductivity behaviors of $\boldsymbol{k^*}$ described using the Ashby method [102].

$$k^* = \frac{1}{3}\left(\left(\frac{\rho^*}{\rho_s}\right) + 2\left(\frac{\rho^*}{\rho_s}\right)^{\frac{3}{2}}\right)k_s + \left(1 - \left(\frac{\rho^*}{\rho_s}\right)\right)k_g \qquad (2.7)$$

This first study revealed that strut thickness played a role in the overall time taken to reach saturation temperature and was shown to have a 15% improvement in cooling in air and 70% improvement in liquid cooling. Their second study tested the same thermal MM with a carbon nanotube film heater and was able to demonstrate a 3.26 times higher cooling rate over conventional cooling techniques used in molding applications [106].

2.10 Vibratory Bandgap Materials

Bandgap behavior in a material refers to the frequency range within a material for which the wave propagation does not occur [107]. These MMs can be of any of the four major categories and are typically periodic unidirectional structures with some form of resonators spaced equidistant from one another across the length of the material [108–110]. These materials are often fabricated through conventional means though additive manufacturing has been shown to be a viable alternative for manufacturing these materials more easily and accurately [111, 112]. Mechanically, these materials will resonate either through vibrations or electromagnetically at various frequencies across the length of the MM. The locations at which the MM reaches resonance represent the actual bandgaps as those locations will not allow the wave of that frequency to propagate through the material [113–116]. Figure 2.14 shows an example of a form which a bandgap material can take [112]. The beam model was developed using Timoshenko beam theory and transfer matrix method placing magnetorheological elastomer resonators periodically along the length of the beam. The test results showed that this material creates bandgap, impeding the transmission of vibrations between the frequencies of 50 and 100 Hz.

A unique mechanical MM developed for high-static-low-dynamic-stiffness applications demonstrated a novel band gap material for low frequency applications [117]. This material uses a regulatory mechanism controlled by an electrically

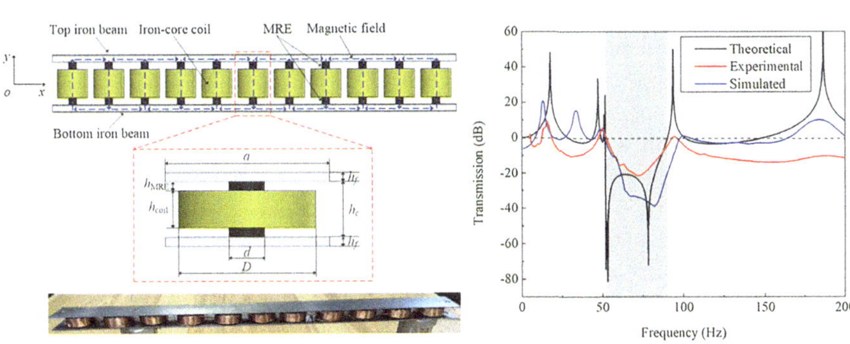

Fig. 2.14 Graphical representation of magnetorheological elastomer sandwich beam with physical beam along with testing data demonstrating the bandgap of the material between 50 and 100 Hz adapted from [109]

charged solenoid to form a high-static-low-dynamic-stiffness resonator to produce a semi-active quasi-zero-stiffness resonator. The material uses a linear array of these quasi-zero-stiffness resonators which can be simplified as simple mass spring systems, where the stiffness in the spring can be adjusted through a magnetic field through an electric load applied to the lower metal plate below the spring mass. To describe the movement of this dynamic system, an equation can be drawn which maps the motion of the system—this is called the equation of motion. The equation of motion of this system is [114]:

$$EI\frac{\partial^4 w(x,t)}{\partial x^4} + \rho S\frac{\partial^2 w(x,t)}{\partial t^2} = 0 \tag{2.8}$$

From here the other equations governing this material can be found. The transmissibility of this material was found numerically across three differing conditions. Four distinct bands were found for which waves would not propagate through the material according to multiple nonlinearity conditions. The semi-active nature of this material allows for a wide range of functions as adjusting the applied voltage changes the stiffness and local resonance at specified locations along the beam, opening various gaps in transmissibility for the material. Depending on the requirement of an application, stiffness and the resonance can be tuned as well as the applied voltage can be used to obtain the desired performance.

2.11 Summary

Physical property manipulation offering control over mechanical behavior of materials is a novel possibility that is exploited in designing metamaterials. The design principles, functionalities, and potential applications of mechanical MM, thermal MMs, and acoustic MMs have been discussed here. Through strategically designing the unit cell geometry and selecting appropriate composition, these materials can be tailored to exhibit a wide range of desired physical properties allowing for the creation of new metamaterials suited for highly specific engineering challenges.

The potential applications of these materials with manipulated physical properties span multiple engineering disciplines such as mechanical, civil, and biomedical engineering. Auxetic materials offer exceptional energy-absorbing capabilities due to their negative Poisson's ratio, making them useful for high-impact protection and vibration dampening. Thermally responsive metamaterials are designed in conjunction with auxetic geometry, altering their stiffness and thermal expansion in response to temperature changes, enabling them to function as self-regulating structures. Studies into buckling behavior and further development of these materials such as understanding bandgap behavior will promise to unlock further applications in the manipulation of physical properties for lightweight, structural, and dynamic applications.

Overall, metamaterials designed for physical property manipulation represent a transformative technology with the potential to revolutionize engineering. Their unique design principles, tunable and tailorable properties, combine with a diverse portfolio of functionalities, open a new avenue for the future where the materials used in an application serve multiple functions to streamline the overall construction of a product.

References

1. Aladwani, A., & Nouh, M. (2020). Mechanics of metadamping in flexural dissipative metamaterials: Analysis and design in frequency and time domains. *International Journal of Mechanical Sciences, 173*, 105459.
2. Porubov, A. V., & Krivtsov, A. M. (2022). Dispersive propagation of localized waves in a mass-in-mass metamaterial lattice. *Continuum Mechanics and Thermodynamics, 34*(6), 1475–1483.
3. Zhang, Y., Cai, J., Mi, C., Wang, F., & Akbarzadeh, A. H. (2022). Effect of surface residual stress and surface layer stiffness on mechanical properties of nanowires. *Acta Mechanica, 233*(1), 233–257.
4. Caeti, R., Gupta, N., & Porfiri, M. (2009). Processing and compressive response of functionally graded composites. *Materials Letters, 63*(22), 1964–1967.
5. Abbasi, T. U. R., & Zheng, H. (2020). Wave dispersion and dissipation performance of locally resonant acoustic metamaterials using an internal variable model. *Wave Motion, 93*, 102483.
6. Lin, Q., Zhou, J., Wang, K., Xu, D., Wen, G., Wang, Q., & Cai, C. (2022). Low-frequency locally resonant band gap of the two-dimensional quasi-zero-stiffness metamaterials. *International Journal of Mechanical Sciences, 222*, 107230.
7. Collet, M., Ouisse, M., Ruzzene, M., & Ichchou, M. N. (2011). Floquet–Bloch decomposition for the computation of dispersion of two-dimensional periodic, damped mechanical systems. *International Journal of Solids and Structures, 48*(20), 2837–2848.
8. Zhao, P., Zhang, K., & Deng, Z. (2020). Elastic wave propagation in lattice metamaterials with Koch fractal. *Acta Mechanica Solida Sinica, 33*(5), 600–611.
9. Arretche, I., & Matlack, K. H. (2023). Centrifugal forces enable band gaps that self-adapt to synchronous vibrations in rotating elastic metamaterial. *Mechanical Systems and Signal Processing, 202*, 110689.
10. Sun, Y., Han, Q., Jiang, T., & Li, C. (2024). Coupled bandgap properties and wave attenuation in the piezoelectric metamaterial beam on periodic elastic foundation. *Applied Mathematical Modelling, 125*, 293–310.
11. Yang, F., Yang, J.-S., Wang, Y., Li, S., Zhang, M.-G., Schmidt, R., & Schröder, K.-U. (2023). Theoretical study on dispersion relations of chiral acoustic metamaterials considering mass-rotation. *European Journal of Mechanics - A/Solids, 100*, 105005.
12. Schiavone, A., & Wang, X. (2021). Continuous modelling of a class of periodic elastic metamaterials with local rotation. *Zeitschrift für Angewandte Mathematik und Physik, 73*(1), 29.
13. Yu, M., Fang, X., & Yu, D. (2021). Combinational design of linear and nonlinear elastic metamaterials. *International Journal of Mechanical Sciences, 199*, 106422.
14. Rokoš, O., Zeman, J., Doškář, M., & Krysl, P. (2020). Reduced integration schemes in micromorphic computational homogenization of elastomeric mechanical metamaterials. *Advanced Modeling and Simulation in Engineering Sciences, 7*(1), 19.
15. Wang, B., Liu, J., Soh, A. K., & Liang, N. (2022). On band gaps of nonlocal acoustic lattice metamaterials: A robust strain gradient model. *Applied Mathematics and Mechanics, 43*(1), 1–20.

16. Fedotovskii, V. C. (2018). A porous medium as an acoustic metamaterial with negative inertial and elastic properties. *Acoustical Physics, 64*(5), 548–554.
17. Zhi, J., Raju, K., Tay, T.-E., & Tan, V. B. C. (2021). Transient multi-scale analysis with micro-inertia effects using Direct FE^2 method. *Computational Mechanics, 67*(6), 1645–1660.
18. Karampour, S., Ghavanloo, E., & Fazelzadeh, S. A. (2023). Free vibration analysis of elastic metamaterial circular curved beams with locally resonant microstructures. *Archive of Applied Mechanics, 93*(1), 323–333.
19. He, Z.-H., Wang, Y.-Z., & Wang, Y.-S. (2021). Active feedback control of sound radiation in elastic wave metamaterials immersed in water with fluid–solid coupling. *Acta Mechanica Sinica, 37*(5), 803–825.
20. Yoon, J. Y., Lim, S., Yoo, J., & Park, N.-C. (2022). Vibration reduction of cables with pendulum-type elastic metamaterials. *International Journal of Mechanical Sciences, 220*, 107169.
21. Qian, Y.-J., Cui, Q.-D., Yang, X.-D., & Zhang, W. (2020). Manipulating transverse waves through 1D metamaterial by longitudinal vibrations. *International Journal of Mechanical Sciences, 168*, 105296.
22. Sepehri, S., Jafari, H., Mosavi Mashhadi, M., Hairi Yazdi, M. R., & Seyyed Fakhrabadi, M. M. (2020). Tunable elastic wave propagation in planar functionally graded metamaterials. *Acta Mechanica, 231*(8), 3363–3385.
23. Shaogang, L., Zhao, Y., Zhao, D., Wu, J., & Gao, C. (2020). Tunable elastic wave bandgaps and waveguides by acoustic metamaterials with magnetorheological elastomer. *Acoustical Physics, 66*(2), 123–131.
24. Zhang, X., Li, K., Li, L., Gao, X., Zhang, J., Hu, N., & Zhang, C. (2023). Temperature-controlled tunable zigzag metamaterial beams to manipulate flexural waves. *Journal of Sound and Vibration, 563*, 117863.
25. Poggetto, V. F. D., & Serpa, A. L. (2020). Elastic wave band gaps in a three-dimensional periodic metamaterial using the plane wave expansion method. *International Journal of Mechanical Sciences, 184*, 105841.
26. Xu, X., Barnhart, M. V., Fang, X., Wen, J., Chen, Y., & Huang, G. (2019). A nonlinear dissipative elastic metamaterial for broadband wave mitigation. *International Journal of Mechanical Sciences, 164*, 105159.
27. Settimi, V., Lepidi, M., & Bacigalupo, A. (2021). Nonlinear dispersion properties of one-dimensional mechanical metamaterials with inertia amplification. *International Journal of Mechanical Sciences, 201*, 106461.
28. Torabi, J., & Niiranen, J. (2021). Microarchitecture-dependent nonlinear bending analysis for cellular plates with prismatic corrugated cores via an anisotropic strain gradient plate theory of first-order shear deformation. *Engineering Structures, 236*, 112117.
29. Wang, Y.-C., Ko, C.-C., Chang, K.-W., & Ko, T.-W. (2021). Negative-stiffness composite systems and their coupled-field properties. *Continuum Mechanics and Thermodynamics, 33*(4), 1857–1872.
30. Klein, J. T., & Karpov, E. G. (2020). Exact analytical solutions in two dimensional plate-like mechanical metamaterials: State of free deformation in a topological cylinder. *International Journal of Mechanical Sciences, 167*, 105292.
31. Reid, D. R., Pashine, N., Wozniak, J. M., Jaeger, H. M., Liu, A. J., Nagel, S. R., & de Pablo, J. J. (2018). Auxetic metamaterials from disordered networks. *Proceedings of the National Academy of Sciences, 115*(7), E1384–E1390.
32. Shen, M., Reyes-Martinez, M. A., Powell, L. A., Iadicola, M. A., Sharma, A., Byléhn, F., Pashine, N., Chan, E. P., Soles, C. L., Jaeger, H. M., & de Pablo, J. J. (2024). An autonomous design algorithm to experimentally realize three-dimensionally isotropic auxetic network structures without compromising density. *npj Computational Materials, 10*(1), 113.
33. Schwarz, D., Felsch, G., Tauber, F., Schiller, S., & Slesarenko, V. (2023). Exploiting self-contact in mechanical metamaterials for new discrete functionalities. *Materials & Design, 236*, 112468.

34. Wang, L., Tan, X., Zhu, S., Wang, B., Li, S., Zou, Y., & Chen, S. (2021). Directional instability-driven strain-dependent 3D auxetic metamaterials. *International Journal of Mechanical Sciences, 199*, 106408.
35. Li, T., & Li, Y. (2022). Effects of handedness ratio and handedness distribution on the mechanical properties of Bi-chiral mechanical metamaterials. *International Journal of Solids and Structures, 254-255*, 111947.
36. Sheykhi, M. (2023). Vibroacoustic performance of star-shaped honeycomb-core annular cellular structures. *Archive of Applied Mechanics, 93*(2), 841–860.
37. Chen, M., Jiang, H., Zhang, H., Li, D., & Wang, Y. (2018). Design of an acoustic superlens using single-phase metamaterials with a star-shaped lattice structure. *Scientific Reports, 8*(1), 1861.
38. Li, Q., & Yang, D. (2020). Vibro-acoustic performance and design of annular cellular structures with graded auxetic mechanical metamaterials. *Journal of Sound and Vibration, 466*, 115038.
39. Spagnuolo, M., & Cazzani, A. M. (2021). Contact interactions in complex fibrous metamaterials. *Continuum Mechanics and Thermodynamics, 33*(4), 1873–1889.
40. Erden Yildizdag, M., Placidi, L., & Turco, E. (2023). Modeling and numerical investigation of damage behavior in pantographic layers using a hemivariational formulation adapted for a Hencky-type discrete model. *Continuum Mechanics and Thermodynamics, 35*(4), 1481–1494.
41. Tang, Z., Ma, T., Li, S., Chen, H., Su, B., Kang, P., Wang, J., Li, P., Wu, B., Qian, Z., & Shi, H. (2023). Polarization-dependent boundary modes in nonlinear mechanical metamaterials. *International Journal of Mechanical Sciences, 258*, 108584.
42. Zhou, Y., Ye, L., & Chen, W. (2022). Impact mitigation performance of hybrid metamaterial with a low frequency bandgap. *International Journal of Mechanical Sciences, 213*, 106863.
43. Zhang, Q., & Sun, Y. (2023). Novel metamaterial structures with negative thermal expansion and tunable mechanical properties. *International Journal of Mechanical Sciences, 261*, 108692.
44. Xu, W., Zhang, L., Zhang, B., Zhang, H., Liu, Z., & Zhu, P. (2023). Crushing behavior of contact-aided AlSi10Mg sandwich structure based on chiral mechanical metamaterials. *International Journal of Mechanical Sciences, 260*, 108636.
45. Ma, J., Song, J., & Chen, Y. (2018). An origami-inspired structure with graded stiffness. *International Journal of Mechanical Sciences, 136*, 134–142.
46. Montazeri, A., Naderinejad, M., Mahnama, M., & Hasani, A. (2023). 3D-printed twisting tubular metamaterials with tunable mechanical and torsional characteristics. *International Journal of Mechanical Sciences, 262*, 108719.
47. Mo, S., Chen, K., Tang, X., & Zhang, W. (2023). Metamaterial design and continuously adjustable mechanical properties based on novel planetary gear trains. *Structure, 58*, 105466.
48. Zhu, Y., Jiang, S., Lu, F., & Ren, X. (2022). A novel enhanced anti-tetra-missing rib auxetic structure with tailorable in-plane mechanical properties. *Engineering Structures, 262*, 114399.
49. Fu, Y., & Liu, W. (2021). Design of mechanical metamaterial with controllable stiffness using curved beam unit cells. *Composite Structures, 258*, 113195.
50. Huang, W., Mei, H., Chang, P., Pan, L., Cheng, L., & Zhang, L. (2022). Tailoring soft-flexible negative Poisson's ratio support to boost photocatalytic efficiency: From dissociation to anchoring. *Additive Manufacturing, 58*, 103059.
51. Yi, J., Zhang, Y., & Chen, C. Q. (2023). Tunable mode conversion in a mechanical metamaterial via second harmonic generation. *Journal of Sound and Vibration, 565*, 117911.
52. Yin, X., Gao, Z.-Y., Zhang, S., Zhang, L.-Y., & Xu, G.-K. (2020). Truncated regular octahedral tensegrity-based mechanical metamaterial with tunable and programmable Poisson's ratio. *International Journal of Mechanical Sciences, 167*, 105285.
53. Montazeri, A., Bahmanpour, E., & Safarabadi, M. (2023). A Poisson's ratio sign-switching mechanical metamaterial with tunable stiffness. *International Journal of Mechanical Sciences, 260*, 108670.

54. Yang, H., Wang, B., & Ma, L. (2019). Designing hierarchical metamaterials by topology analysis with tailored Poisson's ratio and Young's modulus. *Composite Structures, 214*, 359–378.
55. Jiao, P., Chen, T., & Xie, Y. (2021). Self-adaptive mechanical metamaterials (SMM) using shape memory polymers for programmable postbuckling under thermal excitations. *Composite Structures, 256*, 113053.
56. Behravan-Rad, A., & Jafari, M. (2020). Hygroelasticity analysis of an elastically restrained functionally graded porous metamaterial circular plate resting on an auxetic material circular plate. *Applied Mathematics and Mechanics, 41*(9), 1359–1380.
57. Su, Y., Xu, X., Shi, J., & Huang, G. (2021). A 3D mechanism-driven hexagonal metamaterial: Evaluation of Auxetic behavior. *International Journal of Mechanical Sciences, 209*, 106699.
58. Li, C., Shen, H.-S., & Wang, H. (2020). Postbuckling behavior of sandwich plates with functionally graded auxetic 3D lattice core. *Composite Structures, 237*, 111894.
59. Ameen, M. M., Rokoš, O., Peerlings, R. H. J., & Geers, M. G. D. (2018). Size effects in nonlinear periodic materials exhibiting reversible pattern transformations. *Mechanics of Materials, 124*, 55–70.
60. Lin, G., Li, J., Chen, P., Sun, W., Chizhik, S. A., Makhaniok, A. A., Melnikova, G. B., & Kuznetsova, T. A. (2021). Buckling of lattice columns made from three-dimensional chiral mechanical metamaterials. *International Journal of Mechanical Sciences, 194*, 106208.
61. Li, Q., Yang, D., Ren, C., & Mao, X. (2022). A systematic group of multidirectional buckling-based negative stiffness metamaterials. *International Journal of Mechanical Sciences, 232*, 107611.
62. Hao, J., Han, D., Zhang, X. G., Zhang, Y., Jiang, W., Teng, X. C., Lang, J. P., Pan, Y., Ni, X. H., Zhang, X. Y., Xie, Y. M., & Ren, X. (2022). Novel dual-platform lightweight metamaterials with auxeticity. *Engineering Structures, 270*, 114891.
63. Ren, C., Li, Q., & Yang, D. (2020). Quasi-static and sound insulation performance of a multifunctional cylindrical cellular shell with bidirectional negative-stiffness metamaterial cores. *International Journal of Mechanical Sciences, 180*, 105662.
64. Teng, X. C., Ren, X., Zhang, Y., Jiang, W., Pan, Y., Zhang, X. G., Zhang, X. Y., & Xie, Y. M. (2022). A simple 3D re-entrant auxetic metamaterial with enhanced energy absorption. *International Journal of Mechanical Sciences, 229*, 107524.
65. Lyu, S., Qin, B., Deng, H., & Ding, X. (2021). Origami-based cellular mechanical metamaterials with tunable Poisson's ratio: Construction and analysis. *International Journal of Mechanical Sciences, 212*, 106791.
66. Sun, Y., Song, K., Ju, J., & Zhou, X. (2023). Curved-creased origami mechanical metamaterials with programmable stabilities and stiffnesses. *International Journal of Mechanical Sciences*, 108729.
67. Wei, Z., Hu, Z., Zhu, R., Chen, Y., & Hu, G. (2023). A transformable anisotropic 3D pentamode metamaterial. *Materials & Design, 234*, 112306.
68. Li, Z., Yang, Q., Fang, R., Chen, W., & Hao, H. (2021). Origami metamaterial with two-stage programmable compressive strength under quasi-static loading. *International Journal of Mechanical Sciences, 189*, 105987.
69. Meng, Z., Ouyang, Z., & Chen, C. Q. (2021). Multi-step metamaterials with two phases of elastic and plastic deformation. *Composite Structures, 271*, 114152.
70. Meng, Z., Gao, X., Yan, H., Liu, M., Cao, H., Mei, T., & Qing Chen, C. (2024). Cage-shaped self-folding mechanical metamaterials. *International Journal of Solids and Structures, 286-287*, 112560.
71. Zhang, J., Karagiozova, D., You, Z., Chen, Y., & Lu, G. (2019). Quasi-static large deformation compressive behaviour of origami-based metamaterials. *International Journal of Mechanical Sciences, 153-154*, 194–207.

72. Yamaguchi, K., Miyazawa, Y., Yasuda, H., Song, Y., Shimokawa, S., Gandhi, U., & Yang, J. (2023). Post-fabrication tuning of origami-inspired mechanical metamaterials based on Tachi-Miura Polyhedron. *Materials & Design, 233*, 112170.
73. Farzinazar, S., Z. Ren, and J. Lee. *Thermal Conductivity of Graphite Microlattices*. in 2018 *17th IEEE Intersociety Conference on Thermal and Thermomechanical Phenomena in Electronic Systems (ITherm)*. 2018.
74. Uhlířová, T., & Pabst, W. (2019). Thermal conductivity and Young's modulus of cubic-cell metamaterials. *Ceramics International, 45*(1), 954–962.
75. Mirabolghasemi, A., Akbarzadeh, A. H., Rodrigue, D., & Therriault, D. (2019). Thermal conductivity of architected cellular metamaterials. *Acta Materialia, 174*, 61–80.
76. Yan, C., Wang, Z., Qu, J., Chen, Y., & Ma, B. (2023). Scalable and dynamically passive thermal regulation over solar wavelengths enabled by phase-transition metamaterials. *Solar Energy, 257*, 257–265.
77. Li, J., Liu, H.-T., & Zhang, Z.-Y. (2023). Stiffness characteristics for bi-directional tunable thermal expansion metamaterial based on bi-material triangular unit. *International Journal of Mechanical Sciences, 241*, 107983.
78. Shi, C., Mahlmeister, N. H., Luxmoore, I. J., & Nash, G. R. (2018). Metamaterial-based graphene thermal emitter. *Nano Research, 11*(7), 3567–3573.
79. Zhang, Q., Zhu, W., Feng, J., Zhou, J., & Deng, Y. (2021). High-efficient heat flux manipulation of micro-scale thermal metamaterials with facile functional unit design. *Materials & Design, 204*, 109657.
80. Lu, F., Lin, B., Ling, X., Zhang, C., & Zhu, Y. (2023). Controllable design of bi-material metamaterials with programmable thermal expansion and Poisson's ratio. *Composite Structures, 322*, 117417.
81. Xu, M., Zhao, Z., Wang, P., Zhang, Y., Guo, X., Lei, H., & Fang, D. (2022). Planar bi-metallic lattice with tailorable coefficient of thermal expansion. *Acta Mechanica Sinica, 38*(7), 421546.
82. Long, L.-R., Fu, M.-H., & Hu, L.-L. (2021). Novel metamaterials with thermal-torsion and tensile-torsion coupling effects. *Composite Structures, 259*, 113429.
83. Lim, T.-C. (2020). Metacomposite structure with sign-changing coefficients of hygrothermal expansions inspired by Islamic motif. *Composite Structures, 251*, 112660.
84. Peng, Y., Wei, K., Mei, M., Yang, X., & Fang, D. (2021). Simultaneously program thermal expansion and Poisson's ratio in three dimensional mechanical metamaterial. *Composite Structures, 262*, 113365.
85. Zhang, X., Ye, H., Wei, N., Tao, R., & Luo, Z. (2021). Design optimization of multifunctional metamaterials with tunable thermal expansion and phononic bandgap. *Materials & Design, 209*, 109990.
86. Zhang, Y. C., Liang, Y. J., Liu, S. T., & Su, Y. D. (2019). A new design of dual-constituent triangular lattice metamaterial with unbounded thermal expansion. *Acta Mechanica Sinica, 35*(3), 507–517.
87. Chen, J., Xu, W., Wei, Z., Wei, K., & Yang, X. (2021). Stiffness characteristics for a series of lightweight mechanical metamaterials with programmable thermal expansion. *International Journal of Mechanical Sciences, 202-203*, 106527.
88. Chen, J., Wang, H., Wang, K., Wei, Z., Xu, W., & Wei, K. (2022). Mechanical performances and coupling design for the mechanical metamaterials with tailorable thermal expansion. *Mechanics of Materials, 165*, 104176.
89. Novak, N., Krstulović-Opara, L., Ren, Z., & Vesenjak, M. (2020). Mechanical properties of hybrid metamaterial with auxetic chiral cellular structure and silicon filler. *Composite Structures, 234*, 111718.
90. Li, X., Gao, L., Zhou, W., Wang, Y., & Lu, Y. (2019). Novel 2D metamaterials with negative Poisson's ratio and negative thermal expansion. *Extreme Mechanics Letters, 30*, 100498.

91. Ai, L., & Gao, X. L. (2018). Three-dimensional metamaterials with a negative Poisson's ratio and a non-positive coefficient of thermal expansion. *International Journal of Mechanical Sciences, 135*, 101–113.
92. Vasudevan, A., Rodríguez-Martínez, J. A., & Romero, I. (2023). Analysis and design of bistable and thermally reversible metamaterials inspired by shape-memory alloys. *International Journal of Solids and Structures, 275*, 112278.
93. Lim, T.-C. (2023). A metamaterial with sign-programmable thermal expansivity and Poisson's ratio constructed from a hybrid of rotating and non-rotating rigid units. *International Journal of Solids and Structures, 284*, 112510.
94. Liu, S., Li, F., Peng, J., & Zhang, X. (2023). Universal model describing the negative thermal expansion coefficients of bending-type two-dimensional metamaterials with chiral/anti-chiral structures. *Composites Communications, 39*, 101559.
95. He, P., Tao, D., Shen, Y., Liang, Y., Dong, J., Fan, H., & Zhang, Q. (2023). Superelastic bionic graphene/nanoceramic metamaterials with tunable thermo-mechanical performances. *Carbon, 214*, 118357.
96. Wei, K., Xiao, X., Chen, J., Wu, Y., Li, M., & Wang, Z. (2021). Additively manufactured bi-material metamaterial to program a wide range of thermal expansion. *Materials & Design, 198*, 109343.
97. Li, Y., Chen, Y., Li, T., Cao, S., & Wang, L. (2018). Hoberman-sphere-inspired lattice metamaterials with tunable negative thermal expansion. *Composite Structures, 189*, 586–597.
98. Kovács, R. (2024). Heat equations beyond Fourier: From heat waves to thermal metamaterials. *Physics Reports, 1048*, 1–75.
99. Pratap, D., & Anantha Ramakrishna, S. (2021). Nanoporous alumina microtubes for metamaterial and plasmonic applications. *Pramana, 95*(3), 136.
100. Kyeong, D. H., Lee, H., Lee, J., Kim, J., Lee, J., Noh, D., & Choi, W. (2023). Enhanced temperature-responsive functions of thermal metamaterials by interface engineering. *International Journal of Heat and Mass Transfer, 215*, 124509.
101. Abdol-Hamid Owens, C., Wang, Y., Farzinazar, S., Yang, C., Lee, H., & Lee, J. (2023). Tunable thermal transport in 4D printed mechanical metamaterials. *Materials & Design, 231*, 111992.
102. You, J.-H., & Park, K. (2021). Design and additive manufacturing of thermal metamaterial with high thermal resistance and cooling capability. *Additive Manufacturing, 41*, 101947.
103. Daoud, A., Abou El-Khair, M. T., Shenouda, A. Y., Fairouz, F., Mohamed, E., Abdel Aziz, M. E., Yanamandra, K., & Gupta, N. (2022). Aging behavior and thermal expansion of novel Pb alloys reinforced with fly ash cenospheres composite foams. *International Journal of Lightweight Materials and Manufacture, 5*(2), 178–187.
104. Xu, X., & Gupta, N. (2019). Application of radial basis neural network to transform viscoelastic to elastic properties for materials with multiple thermal transitions. *Journal of Materials Science, 54*(11), 8401–8413.
105. Zeltmann, S. E., Chen, B., & Gupta, N. (2018). Thermal expansion and dynamic mechanical analysis of epoxy matrix–borosilicate glass hollow particle syntactic foams. *Journal of Cellular Plastics, 54*(3), 463–481.
106. You, J.-H., Lee, J.-W., Oh, S.-H., & Park, K. (2022). Conformal mold heating and cooling using a carbon nanotube film heater and additively manufactured cellular metamaterial. *International Journal of Precision Engineering and Manufacturing-Green Technology, 9*(6), 1463–1476.
107. Ren, T., Li, F., Chen, Y., Liu, C., & Zhang, C. (2020). Improvement of the band-gap characteristics of active composite laminate metamaterial plates. *Composite Structures, 254*, 112831.
108. Li, Y., Dong, X., Li, H., & Yao, S. (2021). Hybrid multi-resonators elastic metamaterials for broad low-frequency bandgaps. *International Journal of Mechanical Sciences, 202-203*, 106501.

109. Yin, S., Huang, J., Zou, Z., Bui, T. Q., Cong, Y., Yu, T., & Zhang, G. (2024). Isogeometric shape optimization for widening band gaps of periodic composite plates. *European Journal of Mechanics - A/Solids, 103*, 105142.
110. Zeighami, F., Palermo, A., & Marzani, A. (2021). Rayleigh waves in locally resonant metamaterials. *International Journal of Mechanical Sciences, 195*, 106250.
111. Vo, N. H., Pham, T. M., Hao, H., Bi, K., & Chen, W. (2022). A reinvestigation of the spring-mass model for metamaterial bandgap prediction. *International Journal of Mechanical Sciences, 221*, 107219.
112. Wang, Y., Yang, J., Chen, Z., Gong, X., Du, H., Zhang, S., Li, W., & Sun, S. (2022). Investigation of a novel MRE metamaterial sandwich beam with real-time tunable band gap characteristics. *Journal of Sound and Vibration, 527*, 116870.
113. Arfan, M., Alkanhal, M. A. S., Ghaffar, A., & Alqahtani, A. H. (2022). Scattering of Laguerre–Gaussian beam from a chiral-coated perfect electromagnetic conductor (PEMC) cylinder. *Journal of Computational Electronics, 21*(1), 253–262.
114. Pranno, A., Greco, F., Leonetti, L., Lonetti, P., Luciano, R., & De Maio, U. (2022). Band gap tuning through microscopic instabilities of compressively loaded lightened nacre-like composite metamaterials. *Composite Structures, 282*, 115032.
115. Baseri, A., & Keshavarz, A. (2022). Designing a double-negative metamaterial photonic crystal using the Thue–Morse sequence. *Journal of Computational Electronics, 21*(1), 270–279.
116. Li, W., Meng, Q., Huang, R., Zhong, Z., & Zhang, B. (2018). Thermally tunable broadband terahertz metamaterials with negative refractive index. *Optics Communications, 412*, 85–89.
117. Wang, K., Zhou, J., Ouyang, H., Cheng, L., & Xu, D. (2020). A semi-active metamaterial beam with electromagnetic quasi-zero-stiffness resonators for ultralow-frequency band gap tuning. *International Journal of Mechanical Sciences, 176*, 105548.

Chapter 3
Energy Manipulation

List of Symbols

α_s	Sound absorbing coefficient
$\boldsymbol{b}$	Radiator periods
$\boldsymbol{c}$	Speed of light
$\boldsymbol{D\ (\Theta, \alpha, \omega)}$	Directivity of sound radiation
$\boldsymbol{Ei}$	Electric field
$\in$	Permittivity
ε_{eff}	Effective permittivity
γ_D	Medium-dependent oscillation frequency
$\hbar$	Plank's constant
$\boldsymbol{k}$	Wave number
$\boldsymbol{K_b}$	Wave vector in Bernoulli zone
$\boldsymbol{K_e}$	Bulk modulus
$\boldsymbol{m}$	Mass
$\boldsymbol{n}$	Mode number
$\boldsymbol{N}$	Radiating cell
$\boldsymbol{ne}$	Electron density
$\boldsymbol{P_i}$	Energy absorbed per unit volume
$\boldsymbol{q}$	Charge
$\boldsymbol{U}$	Inter-band contribution
$w\ (\alpha, \omega)$	Radiated sound pressure interference
z	Impedance
$\boldsymbol{\Gamma}$	Scattering rate
$\boldsymbol{\mu_c}$	Chemical potential of graphene
$\boldsymbol{\rho_e}$	Effective density
$\boldsymbol{\sigma_d}$	Frequency dependent electrical conductivity Tensor
$\boldsymbol{\omega}$	Angular frequency
$\boldsymbol{\omega_p}$	Oscillation frequency
$\boldsymbol{\sigma}$	Electrical conductivity

N. Gupta, C. Beckwith, *Architected Metamaterials: Design Principles and Properties*, https://doi.org/10.1007/978-3-031-98330-6_3

3.1 Introduction

Energy absorbing metamaterials likely have attracted the greatest interest compared to any other class of metamaterials (MMs). These MMs are designed for a broad range of applications using lattice formations ranging from two-dimensional and periodic to three-dimensional and nonperiodic [1]. The energy absorbing metamaterials can be classified as electromagnetic MMs, mechanical MMs, thermal MMs, and acoustic MMs. Each of these material types use properties intrinsic to the waves they are intended to work with to harvest energy. Acoustic MMs absorb energy through transverse and longitudinal acoustic waves as well as vibrations in order to convert them into piezoelectric signals [2]. Similarly electromagnetic MMs convert electromagnetic waves such as radio, infrared, and ultraviolet light into electric signals through Faraday's law and Maxwell's equations to manipulate the electromagnetic fields produced by those waves as electric signals [3]. Mechanical MMs store energy by applying a force to the material, deforming it and releasing that energy by relaxing the force [4]. Lastly, thermal MMs convert thermal waves produced through convection and radiation to electrical signals by means of piezoelectric conversion [5].

Figure 3.1 shows the Ashby chart of specific energy absorption (SEA) vs. density for mechanical MMs [6]. A general trend emerged from this data for SEA and density. Most materials show increasing SEA with density; however, carbon fiber reinforced polymer (CFRP) honeycomb composites retain an almost constant SEA

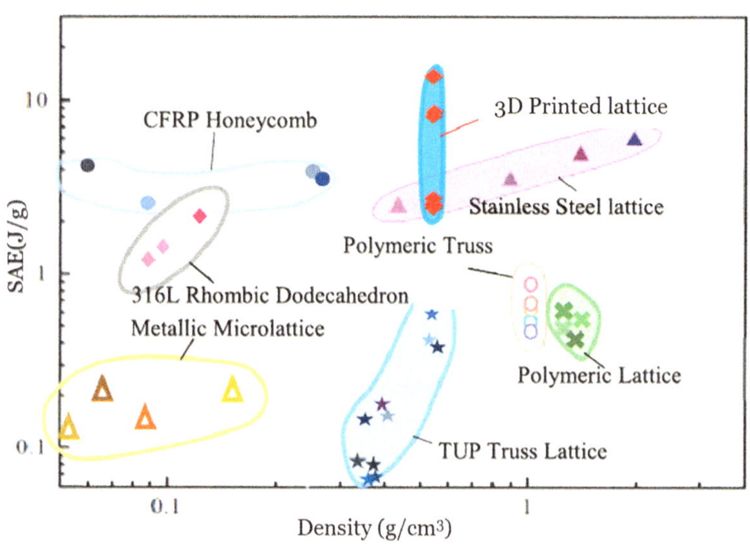

Fig. 3.1 Ashby plot of energy absorbing mechanical MMs according to specific energy absorption vs. density adapted from [6]

regardless of density. In comparison, 3D printed materials exhibit a wider range of SEA within a narrow range of densities based on their microstructure. These trend-breaking lattice structures are ripe for study for their applications such as underwater structures that require low density but high compressive strength.

Figure 3.2a shows the compressive strength of a 3D printed mechanical MM [6], while Fig. 3.2b shows the absorption performance of ceramic composite

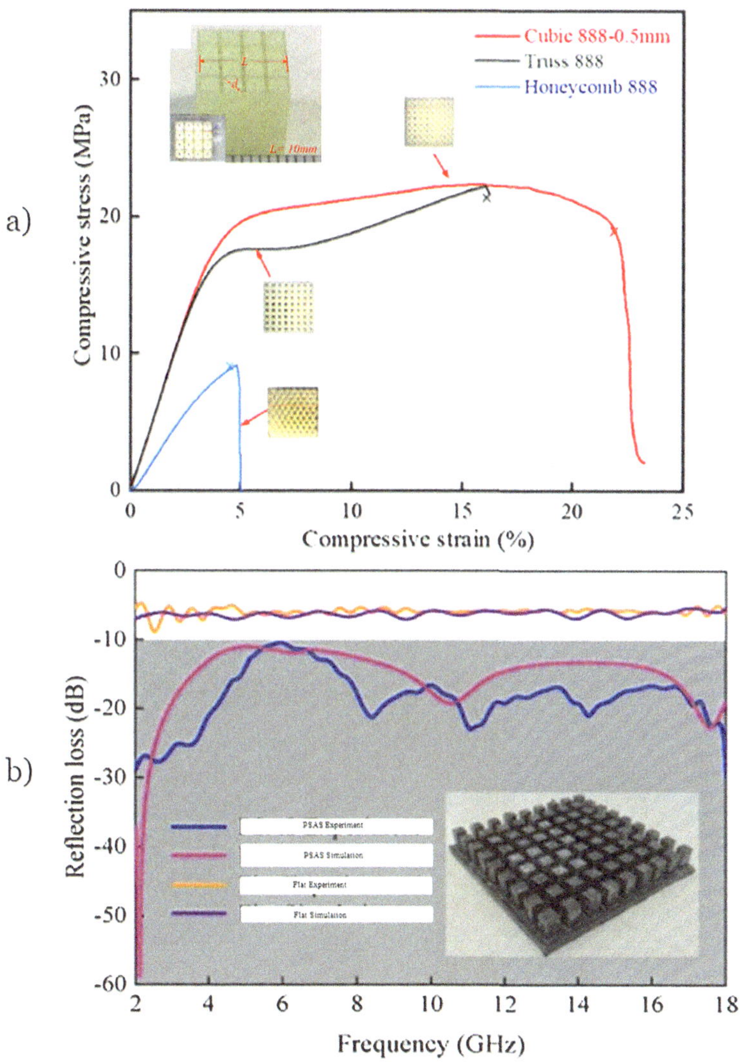

Fig. 3.2 Energy absorbing metamaterials consisting of mechanical, thermal, electromagnetic, and acoustic MM. (**a**) Mechanical MM created from 3D printed resin validated for its specific energy absorption and density adapted from [6]. (**b**) Electromagnetic MM validated for reflection loss at multiple frequencies adapted from [7]

electromagnetic MM in a range of 2–18 gigahertz (GHz) [7]. Figure 3.3 exemplifies the two other types of metamaterials as (a) demonstrated the energy absorbed by a thermal MM in terms of the radiation intensity at multiple wavelengths [8], and (b) contains the experimental set up and performance of an acoustic MM showing an overall higher absorption coefficient than the simulation results of the same material [9].

The manipulation of energy using metamaterials is an important practice as these materials offer an important solution for increasing efficiency in systems that lose energy due to heat and sound. Thermal MMs and acoustic MMs convert energy back into usable electrical energy through sophisticated integrated thermal photovoltaics and piezoelectric absorbers. This opens the potential for high-efficiency

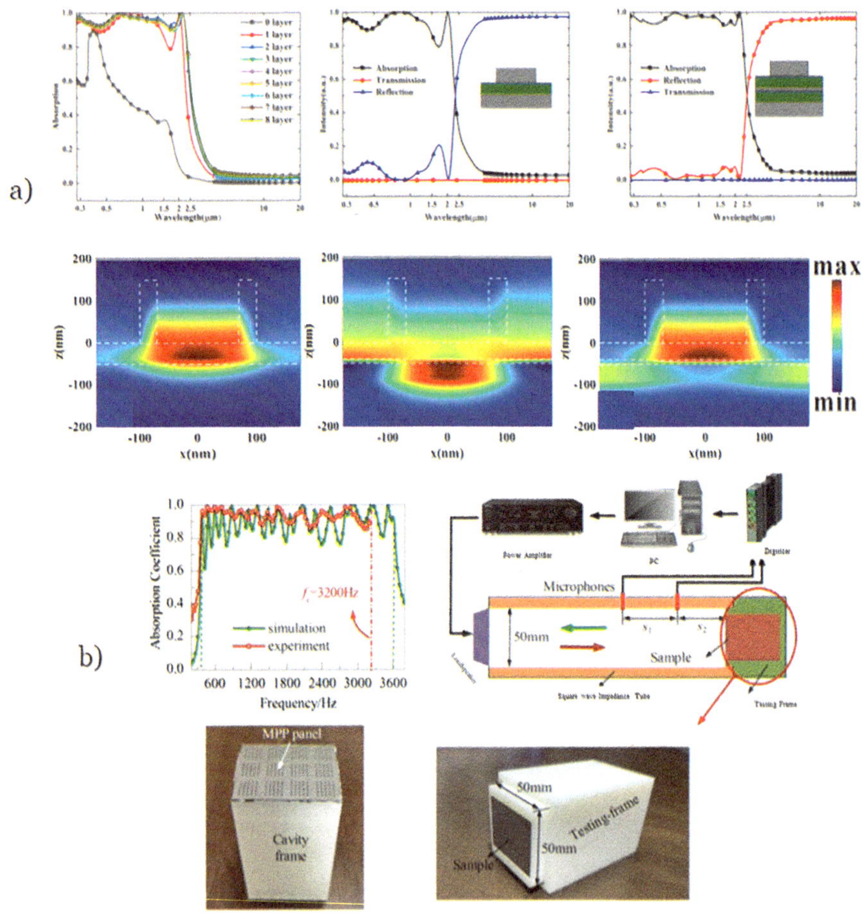

Fig. 3.3 (**a**) Energy absorbing thermal MM validated for wavelength and intensity adapted from [8]. (**b**) Experimental set up for acoustic MM evaluated for absorption at multiple frequencies adapted from [9]

systems, and new possibilities for energy collection. Electromagnetic and thermal absorption methods work similarly where the stimulus passes through the material and either the permittivity or the conductance of the material transforms. That external energy either from radiation of the electromagnetic wave or the heat from an outside source is converted to energy internal to the MM. Mechanical and acoustic absorption also work similarly where the wave or force causes deformation in the material and is converted into internal energy using phenomena like the Helmholtz effect or through dampening in the system to stop the motion caused by the stimulus. The manipulation of energy can be applied to the manipulation of signals and create more robust communications technologies. The manipulation of signals such as sounds and electromagnetic communications are a driving force for MM development creating terahertz (THz) based antennas capable of receiving and transmitting a wide range of signals simultaneously. Another use is the creation of asymmetric transmission materials where signals can only travel in one direction likened to an acoustic diode. These applications demonstrate the effectiveness of metamaterials to enable and enhance fields in ways previously thought impossible.

3.2 Electromagnetic Absorption

Electromagnetic absorption can be accomplished through an exercise of Maxwell's equations simply realizing that electric and magnetic fields change over time and are directly manipulatable through a material's intrinsic properties. The composition and geometric formation of the material dictates the boundary conditions for the waves the material can interact with [10, 11]. These materials are often simple two-dimensional layered structures fabricated through electrochemical means composed of graphene, copper, silicon, and aluminum [12, 13]. The creation of a perfect EM absorber is often the goal of these materials, particularly for sensitive applications such as medical imaging, communications, and stealth [14–16]. Perfect absorbers are designed for their ability to absorb all the EM waves within a given band. These absorbers are developed to interact with gigahertz and terahertz range such that the most common communications bands are accounted for. Metamaterials capable of absorbing between 98.9% and 99.9% of incident EM waves have been achieved, which is very close to the desired capability of 100% absorption [17–21]. Making these materials smaller and more compact creates opportunities for many future applications that have weight or volume limitations, such as satellites or aerospace applications. Another area of focus for electromagnetic MM absorbers revolves around creating thin film-like resonators capable of interacting with infrared and visible light waves. These films are intended for imaging and detection, where the key in fabricating these materials lies in changing the thickness of the unit cell to ensure the permeability and permittivity of the material are as close to 0 as possible [22–24].

Graphene has been a promising material used in composite material development for its high electrical conductivity, broadband absorption, and material compatibility [25]. Electromagnetic MMs utilize this material to enable EM absorption capabilities. Shielding from electromagnetic waves is one method of developing EM absorbing metamaterials. All wave interactions have three components namely absorption, reflection, and transmission [26]. For electromagnetic MMs that absorb and shield against EM waves, the goal is to minimize the transmitted waves as much as possible. Novel metamaterials have been developed in the past that take advantage of graphene's natural EM wave absorption. Graphene combined with hollow particles in a matrix create foamed metamaterials capable of absorbing EM waves, while being very light-weight, suitable for underwater or aerospace stealth applications. Figure 3.4 shows a schematic depiction of this process in (a) demonstrating the basic mechanisms materials deploy when interacting with an incident

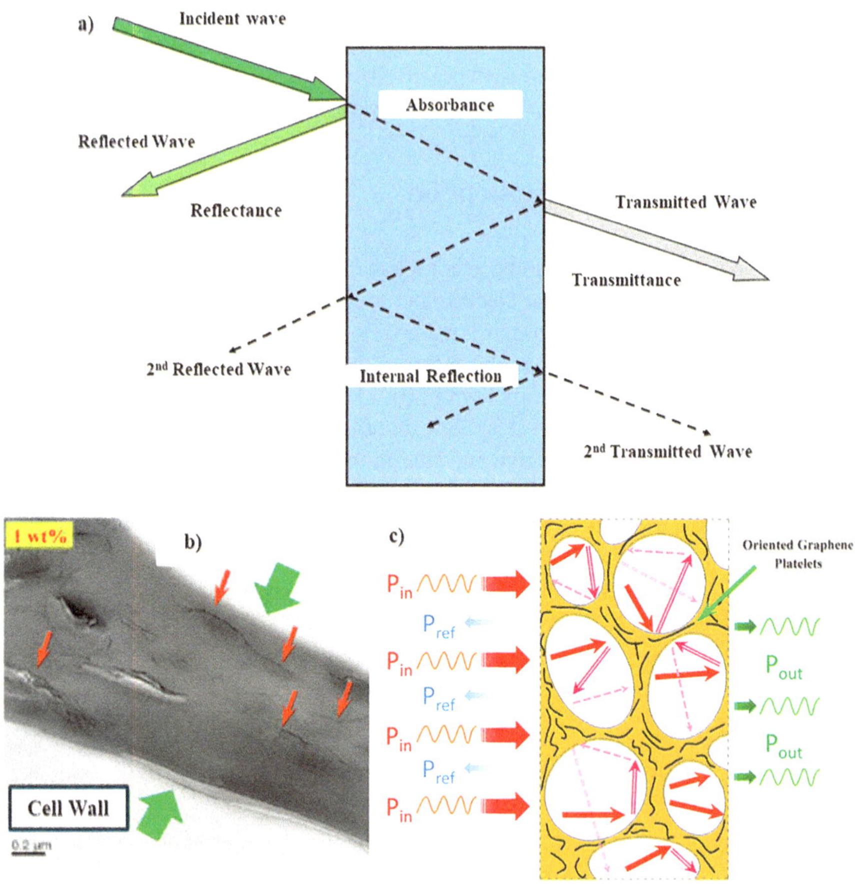

Fig. 3.4 Schematic of the absorption process: (**a**) schematic depicting an incident wave entering a material and the mechanism leading to electromagnetic interference shielding, (**b**) TEM image of the dispersion of graphene in a polyetherimide-graphene nanocomposite foam, and (**c**) schematic representation of microwave transfer across the nanocomposite adapted from [27]

EM wave. Figure 3.4b, c provides a transmission electron microscopy image of the dispersion of graphene in a polyetherimide-graphene nanocomposite foam represented as the small strands with red arrows, along with a schematic representation of the shielding effect that takes place in the material. In this case, when the microwave encounters the composite, the waves are absorbed through the graphene interface and become trapped within the hollow particles inside of the polyetherimide matrix. This leads to a much lower reflection and transmission of microwaves as opposed to a composite without the graphene enhancing the absorptivity of the material [27].

Ceramics and composites play a vital role in the creation of electromagnetic MM absorbers. Graphene, vanadium dioxide, gallium arsenide, and silicon nanofibers have such high properties relevant to absorbing and reflecting EM waves that many electromagnetic MM absorbers implement some combination of these materials to the composition of the MM [28–31].

Often these materials must be modeled for their permittivity. The Drude model, used in modern physics, is commonly employed as it can describe the conductive material's resistance in terms of the scattering of electrons. Maxwell-Garnett approximations, and other effective medium theories, see similar use as they more directly describe the permittivity of an effective medium in terms of the permittivities and volume fractions of individual constituents in an overall medium. The Drude model or Maxwell-Garnett approximations are commonly used to characterize the dielectric constant and permittivity of the material such as [32]:

$$\in(\omega) = 1 - \frac{\omega_p^2}{\omega^2 - i\omega\gamma_D} \tag{3.1}$$

These materials allow for sub-perfect absorbers to function upward of 80% absorption rate over a wideband frequency range generally between 3–6 GHz and 1–4 THz [33–36]. However, this absorption rate is much lower than the ~99% demonstrated by the previously described materials.

Multilayered electromagnetic MM absorbers are commonly studied in the literature as layering multiple two-dimensional unit cells of differing shapes upon one another creates a broadband effect where multiple bands can be absorbed [37, 38]. Since these materials are layered, studies into the microstructure of these materials revealed the effects these geometries have on the ability to absorb multiple frequencies. Electromagnetic MM absorbers with microarchitecture between 5 and 20 μm are suitable to interact with GHz waves where λ = 1–750 mm and ones with microarchitecture between 5 and 500 nm are more suitable for THz applications where λ = 30–300 μm as per these studies [39, 40].

Understanding the effects of a structure's symmetry on its permittivity allows for the creation of a novel approach to electromagnetic MM absorption. Using a 2D quasicrystal lattice structure with five-fold symmetry to act as a GHz absorber, different diffraction patterns, and absorption profiles can be observed [41]. The sample was additively manufactured from polyurethane impregnated with carbonyl iron

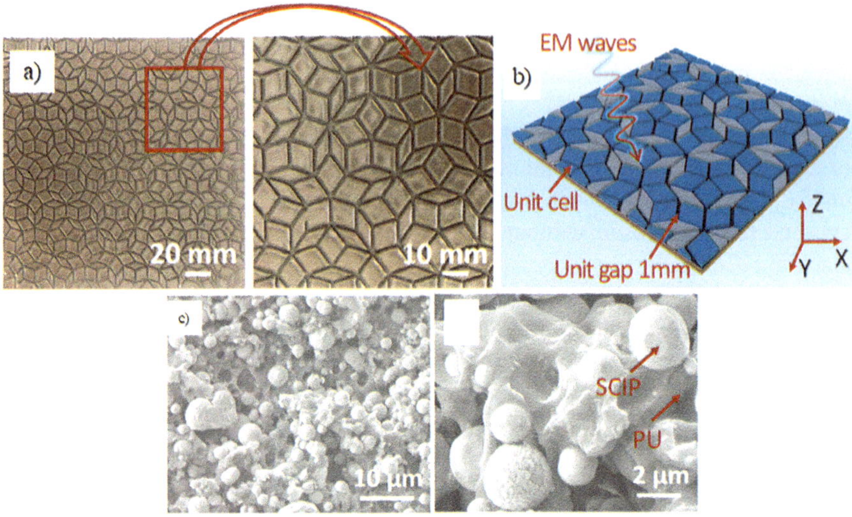

Fig. 3.5 (**a**) Photos of fabricated quasicrystalline electromagnetic MM absorber. (**b**) Schematic diagram with absorption mechanic. (**c**) SEM images of spherical carbonyl iron powder mixed in polyurethane adapted from [40]

powder on top of an aluminum plate. The diffraction pattern found by the nonperiodic quasicrystal structure was shown to produce a higher loss of reflection than a square pattern with four-fold symmetry. This is consistent with the Bragg diffraction condition allowing for a localization of scattering the mean free path. Their experimentation yielded effective absorption within the band of 5.8–18.0 GHz. Figure 3.5 schematically describes the configuration and performance of the quasicrystal lattice structure, showing lower reflection losses than the base slab across multiple configurations. Figure 3.6a provides the analysis of the permittivity and reflection of the quasicrystal structure. Figure 3.6b emphasizes these results by depicting the diffraction pattern of the absorbed light.

3.3 Acoustic Absorption

According to the literature, acoustic MMs encompass the largest category of energy-absorbing MMs. Acoustic MMs typically target two types of acoustics: sound waves and vibrations. Where a sound wave is typically characterized by the vibration of molecules in fluids such as air and water, and vibration waves travel through solids [42, 43]. For sound-based acoustic MM absorbers, noise and sound attenuation are large areas of focus, whereas for vibrations energy harvesting and conversion through piezoelectric processes, and dampening structures are the main focuses. These materials are 2D lattice structures which include unit cells like winding corridors. These act as wave guides not too dissimilar to acoustic cloaks where

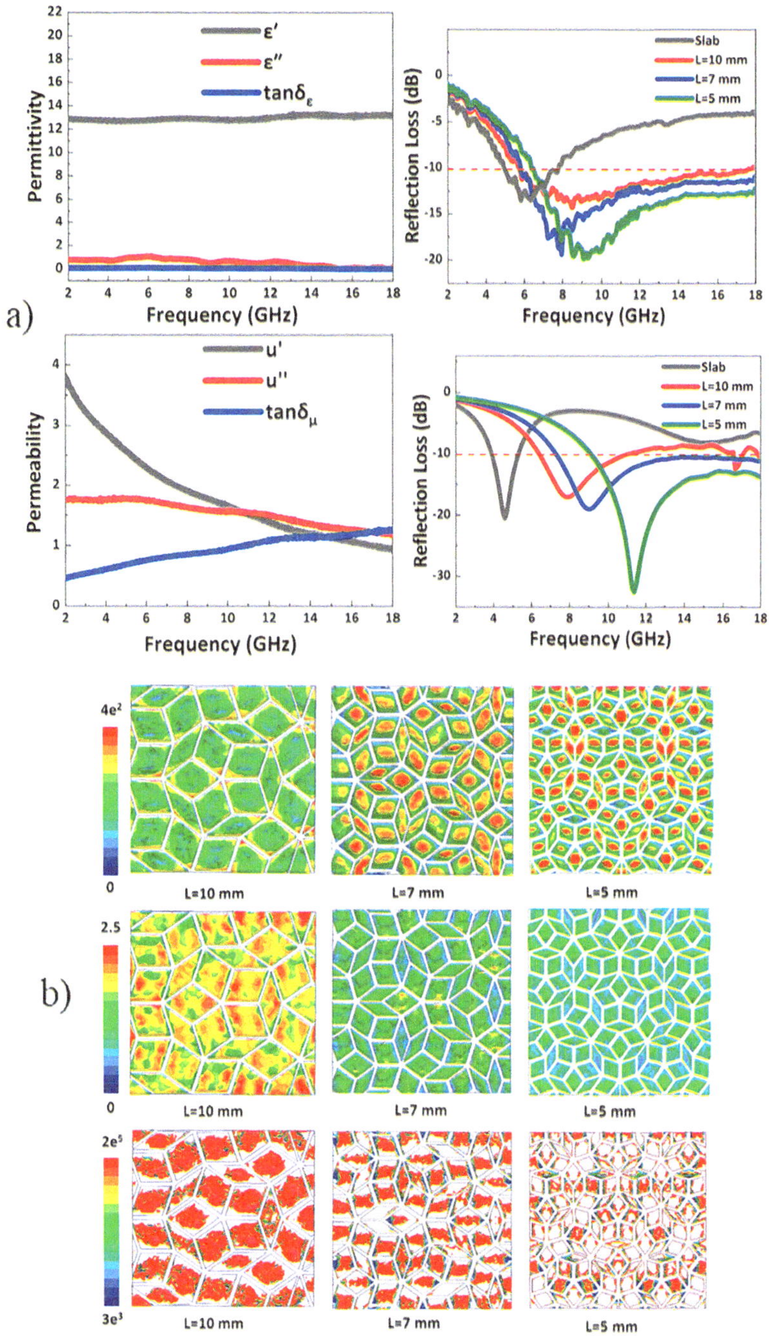

Fig. 3.6 (**a**) Complex permittivity and complex permeability of electromagnetic MM absorber with experimental and simulated reflection loss of raw slab and metamaterial. (**b**) Simulated absorption volume of the lattice structure electric field, magnetic field, and volume loss density distribution adapted from [40]

corridors act as porous channels which guide the sound waves and trap them within [44–47]. Commonly, acoustic MMs convert the mechanical deformation of their unit cells resulting from vibrations into electrical energy [48]. This effect is enabled through Helmholtz resonance. This type of resonance occurs as the sound waves enter a small opening with a completely enclosed chamber. The vibrations inside the chamber resonate with the material the chamber is composed of at its natural frequency causing deformation and inducing the piezoelectric effect [49, 50]. This allows for broadband energy harvesting through the creation of an array of resonators of differing sizes and geometries such that resonance occurs across a large spectrum of frequencies [51, 52].

Vibration suppression is another way acoustic MMs absorb waves. Developed for applications such as damping, structural vibration mitigation, and insulation of dynamic systems [53, 54], these materials are often additively manufactured through material extrusion with acrylonitrile butadiene styrene or polylactic acid but do occasionally use porous materials such as polyimide-impregnated hollow particle filler composites called syntactic foams [55, 56]. Figure 3.7 exemplifies a sound-absorbing acoustic MM created using polyimide aerogel in three configurations demonstrating its sound-absorbing capabilities [52]. Using lattice structures, local resonance can be produced on the material for sound transmission loss which can reduce resonance from 17.7% to 64.4% for low frequency applications [57–60]. When analyzing these materials, simplification into multi-degree-of-freedom systems and examining the lattice as a series of spring mass systems simplifies constructing boundary conditions. This simplifies finding the equation of motion as the unit cell is free to rotate and translate in every direction. As a result, studying nonlinearities on the local level is important through continuum mechanics [61–63].

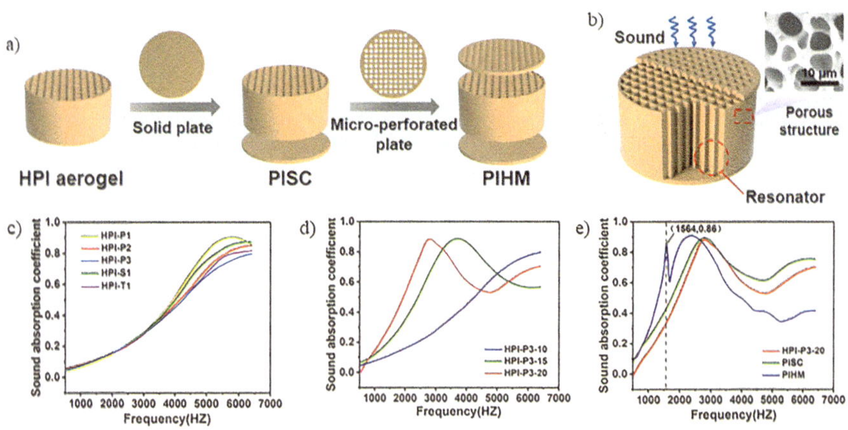

Fig. 3.7 (**a**) Assembly diagram for sound-absorbing acoustic MM. (**b**) Schematic diagram of acoustic MM with SEM microstructure image showing the porosity of the printed aerogel. (**c**) Sound absorption curves of acoustic MM with differing infill porosities and viscosities. (**d**) Sound absorption curves at multiple thicknesses. (**e**) Comparison absorption curves for three absorption structures adapted from [52]

Sound proofing and suppression are remarkably like vibration suppression in theory for acoustic MM absorbers. Stiffness, mass, and dynamic density, impedance, and bulk modulus are important material properties to reduce the sound pressure level within an airspace [64–66]. Here Helmholtz resonance plays a significant role in determining the impedance and sound pressure distribution across the material [67]. This allows for the creation of an impedance tube, an important chamber within which these materials can be validated. The samples evaluated in these tubes are composed of porous materials with stiff sections on either side of the porous material creating a Helmholtz resonator. Within the tube the absorption coefficient can be found over a wide range of frequencies typically between 200 and 1200 Hz.

A Helmholtz resonator describes an air-spring oscillator that is in an enclosed volume having a small neck and an opening at one end. The adaption of this design has important impacts on sound dampening applications for their ability to trap sound within the enclosed volume; this can be described as the sound-absorbing coefficient α_s which can be found numerically through [68]:

$$\alpha_s = 1 - \left| \frac{\mathbf{z} - \mathbf{1}}{z + \mathbf{1}} \right|^2 \tag{3.2}$$

In this case, z is the impedance, or the material's opposition to the wave's flow, across the surface of the material which can be found as the proportion of the effective bulk modulus and effective density [68]:

$$z = \sqrt{\boldsymbol{K}_e \rho_e} \tag{3.3}$$

These materials have been shown experimentally to provide a sound pressure level decrease of up to 21 dB over a broad range of frequencies [69–71].

Aquatic applications are of high importance to acoustic MM studies due to the complexities of developing materials for underwater use. This includes military stealth, sonar, and ultrasonic detection, among other applications. These materials are generally developed as layered plate structures with membrane-like mass spring unit cells periodically distributed throughout the sandwiched plane between the plates [72, 73]. Resistance and continued performance under high hydrostatic pressure are keys to developing these materials for real-world applications. To that end these materials have been shown to produce an average sound transmission loss of up to 10 dB between 2 and 10 kHz under 3 MPa of hydrostatic pressure [74, 75].

3.4 Mechanical Absorption

As discussed with auxetic MMs, energy absorption is a large field of study for mechanical MMs, particularly through the understanding of storing energy in the compression and damping behavior resulting from the lattice structures used in

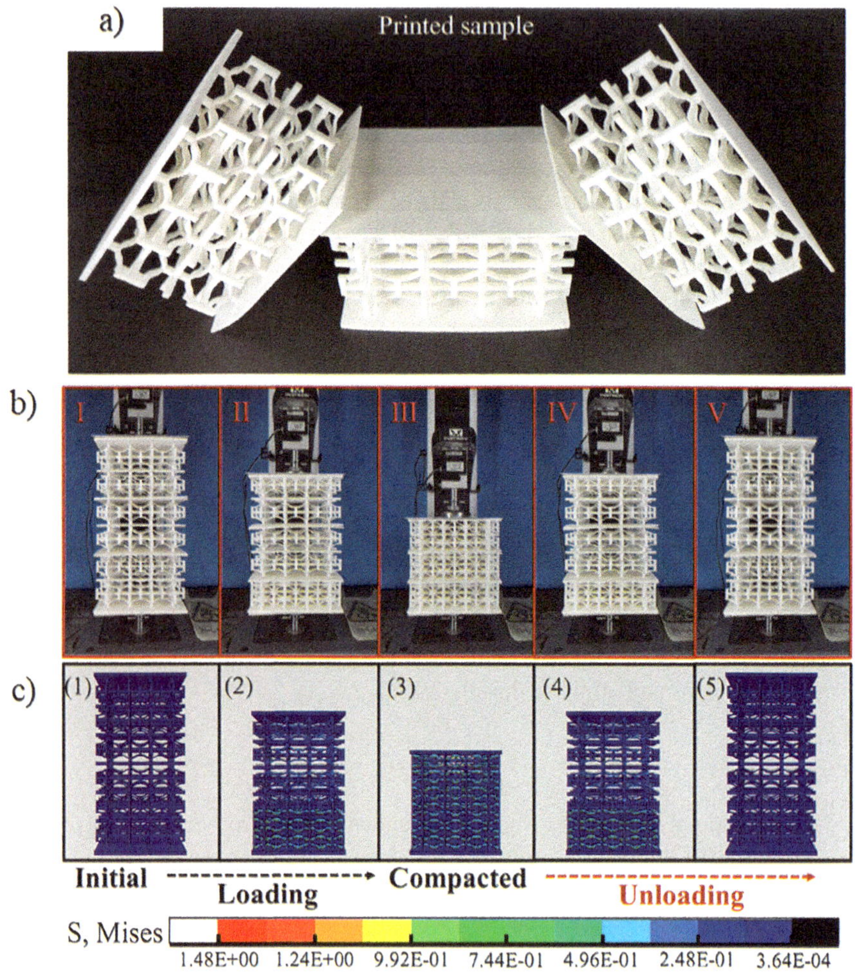

Fig. 3.8 (**a**) Schematic representation of printed metamaterial. (**b**) Dynamic compression testing and simulation of the material at different loaded and unloaded levels (i-v). (**c**) Simulation of compression at different level of loaded and unloaded levels (1–5) adapted from [78]

these materials [4, 76, 77]. These materials are typically developed for energy dissipation, impact resistance, and shock absorbing. Additive manufacturing has helped in validating the simulation results [78, 79]. Figure 3.8 shows thermoplastic polyurethane printed through powder bed fusion and validated through quasistatic loading-unloading and dynamic hammer testing for both the experimental and simulated tests [78]. Figure 3.9 provides the results of these tests for the simulated and experimental results. It can be noted in specimen S3 that the lattice produced an asymmetric buckling pattern which is reflected in its ability to release energy being much lower compared to its ability to dissipate energy. Additively manufactured

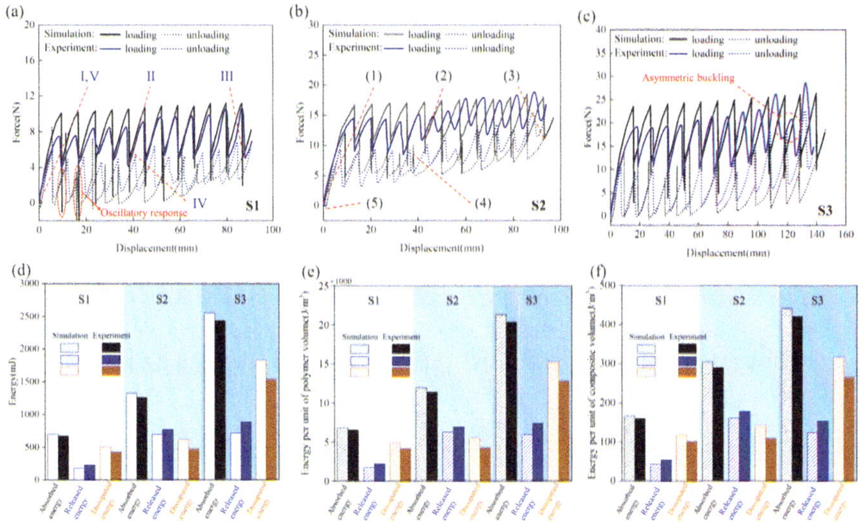

Fig. 3.9 A comparison of load vs. displacement curves of various specimens composed of compressively loaded mechanical MMs for experimental values and simulated results: (**a**) Specimen S1, (**b**) Specimen S2, and (**c**) Specimen S3. (**d**) A comparison of absorbed energy, released energy, and dissipated energy for simulated and experimental values. (**e**) Comparison per unit of polymer volume. (**f**) Comparison per unit of composite volume adapted from [78]

lattices composed of cylinders and chiral lattices are common materials explored for their dissipative properties. Their ability to compress and then form again their previous shape when pressure is removed is the most interesting characteristic [80–82]. Other materials build on this concept creating shape memory lattices, which will recover to an undeformed shape after stopping an impact of a material with an arbitrary shape [83]. In the analysis of these materials, the force and displacement caused by compression can extract the potential and kinetic energy stored in the system and can be mapped over time to show the maximum energy stored within the material [84]. An interesting mechanical MM absorber made the case for reusable meta-structures for inelastic energy absorption [85]. The lattice used a sine wave-like unit cell supported at each crest and trough composed of water jet cut steel that was subject to cyclic loading of up to 1.5 kN to 35 mm. This mechanical MM showed a maximum energy dissipated per unit volume within the megajoules range over 20 cycles.

3.5 Thermal Absorption

Thermal energy absorbing MMs represent an important field of study within material science. The management and control over thermal energy generated in a system is a struggle for engineered materials to overcome. Thermal MMs typically work through several phenomenon such as convection, radiation, or converting the energy through processes such as photo-thermoelectric, piezoelectric, or other dissipative means [8, 86, 87]. These materials can be fabricated by several different methods, most commonly, through electrochemical means such as chemical etching, and mesh screen printing. Their composition typically consists of aluminum, silicon, and gold.

Many thermal metamaterials will work to absorb radiation and be designed to continue operations at high temperatures. To that end modeling these interactions is paramount to understanding their performance after fabrication. Looking at the connection between thermal absorbance due to electromagnetic radiation opens the opportunity for more robust design space. The energy absorbance of the material can be observed through the permittivity of waves allowed through it such as in [88]:

$$P_i = 0.5\left(\omega\epsilon_i + \sigma\right) \times \left|E_i\right|^2 \quad (3.4)$$

where the energy absorbed per unit volume is directly a function of the change of permittivity across a field. Many of these materials are often designed to function at temperatures upward of 800 °C and remain stable well into the terahertz range for frequencies, leading them to be very delicate and precise in construction [5, 88]. Another interesting study detailed their experiments in energy absorbing thermal MMs [89]. The experiments showed a reflectivity between 3–5 μm and 8–14 μm, where the absorption band of high emissivity exists between 5.5 and 7.6 μm. The radiative power of this structure was found experimentally to be 7806 W/m^2 at 600°K. Suggesting these materials would have high potential applications in stealth, imaging, and solar power harvesting, among others. Figure 3.10 illustrates the schematic layout of both an individual and an array of solar absorber metamaterials

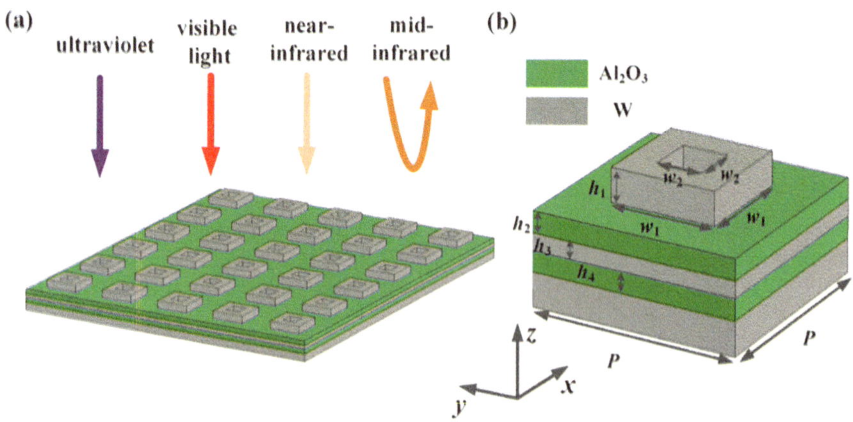

Fig. 3.10 Schematic diagram of thermal absorption metamaterial and corresponding solar absorption spectral distribution: (**a**) Solar irradiance spectrum with solar energy absorbed. (**b**) Solar absorption under thermal electric and thermal magnetic polarizations adapted from [8]

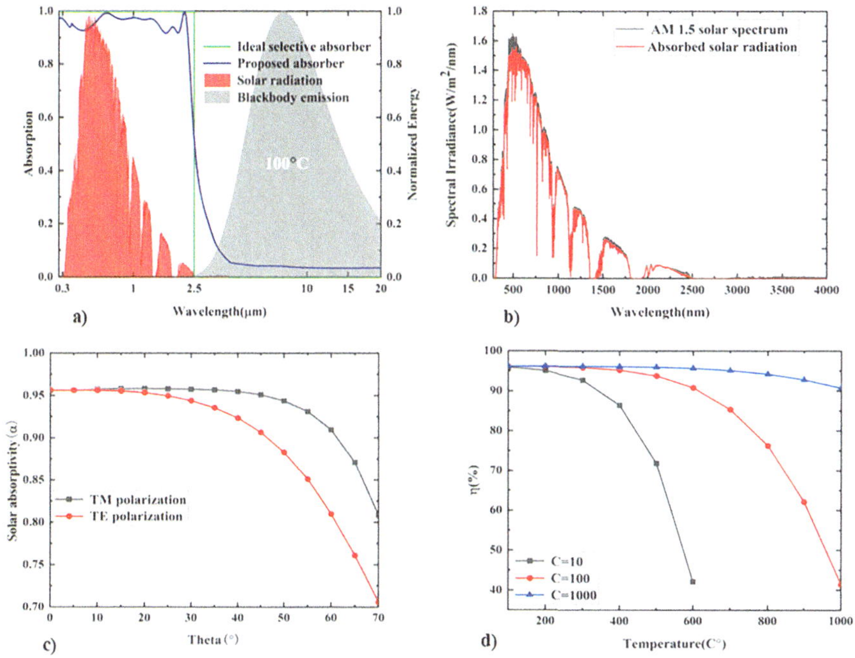

Fig. 3.11 Solar-to-thermal energy conversion efficiencies under different solar concentration factors: (**a**) idealized solar spectrum absorption, (**b**) solar irradiance spectrum, (**c**) solar energy absorbed by the proposed metamaterial under transverse-magnetic and transverse-electric polarization, and (**d**) solar absorption efficiency as a function of temperature adapted from [8]

created from layering tungsten and aluminum oxide. The relationships between the solar energy absorption and conversion of heat are exemplified in Fig. 3.11. These graphs show absorption across the full solar spectrum. Optimal polarization appears at an angle of 45° and high solar thermal conversion at higher temperatures between 100 and 1000 °C [8]. This is an important advancement for metamaterials as the elevated level of solar energy absorption and the few materials used showed potential for developing new solar panels and solar energy harvesters using a wider range of materials and formations that can prove optimal for converting solar energy into thermal power.

Another example of a thermal absorbing thermal MM is depicted in Fig. 3.12 adapted from [88] The schematic depicts an array of high temperature conductive paste laid out periodically on a substrate composed of Al_2O_3 layered on top of a nickel sheet. SEM images of the powder morphology as well as the cross-sectional coating of the Al_2O_3 are provided for quality assurance of the powder, and distribution of the Al_2O_3 plays a significant role in the performance of the metamaterial. The impedance matching can be shown in Fig. 3.13d, e for the single aluminum coating and the proposed metamaterial, respectively, where the metamaterial shows more ideal impedance matching of the real and imaginary parts of the impedance for

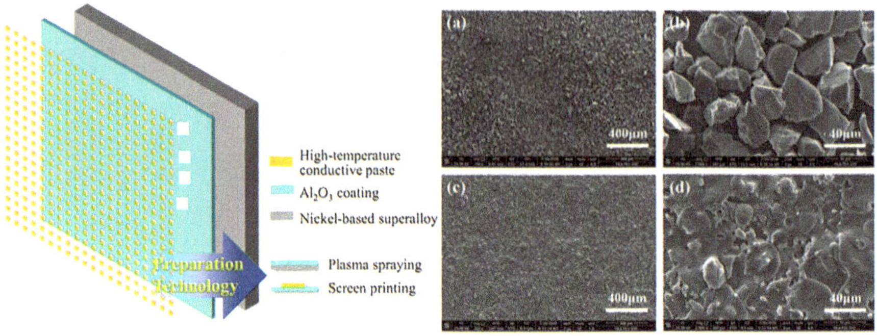

Fig. 3.12 (Left) Schematic diagram of radar absorbing thermal MM. (Right) SEM images of the feedstock powder morphology (**a**, **b**), and cross section of Al_2O_3 coating (**c**, **d**) adapted from [89]

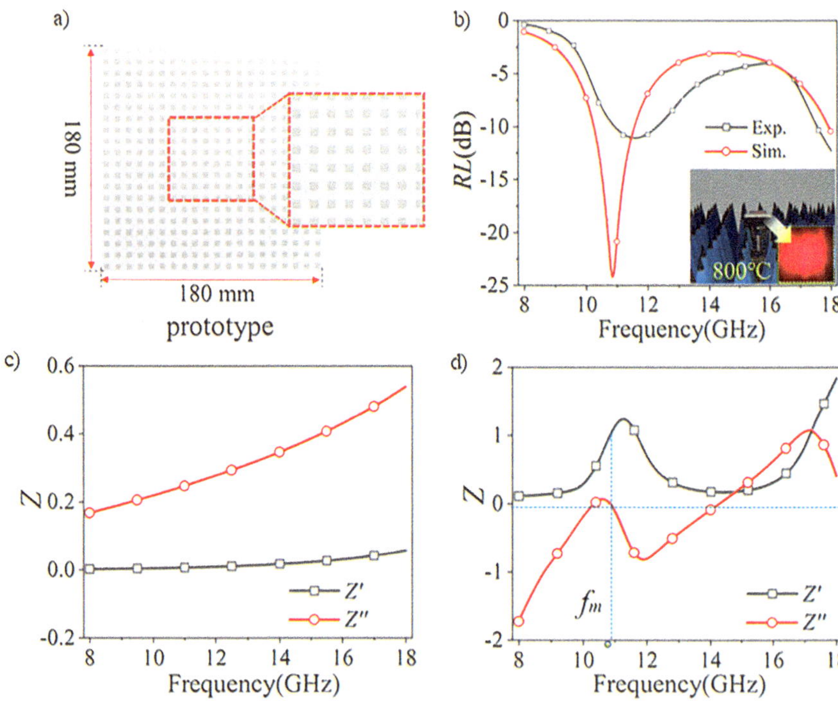

Fig. 3.13 Impedance maps of composite coating (**c**) and thermal MM absorber (**d**) for real (Z′) and imaginary (Z″) portions of the impedance. Lastly (**a**) an image of the prototype of the radar absorber and (**b**) a comparison between the experimental and simulated performance of the absorber evaluated for the reflection loss at 800 °C adapted from [89]

radar absorption at the higher GHz range. The experimental results and simulated results are compared for the reflection loss over the same range of frequencies evaluated at 800 °C. This is also provided with the schematic prototype. While the trend of the two curves is close, the practical performance of the material is noticeably different from the simulated results. Overall, this material showed promise for radar absorption between 8 and 18 GHz at 800 °C; however, this did not show significant improvement over similar materials in terms of performance, but the simplicity in the construction and few materials used were the advantages. The simple materials and construction allow for simple efficient manufacturing and for opportunities to scale this material for broader applications such the detection of a certain spectrum.

3.6 Signals Driven by Metamaterials

Signal driven transmission is a term given to MMs that deal in the transfer of information from one point to another. This includes antennas with THz frequency ranges and perform multi-band data transmission. Additionally, materials which unidirectionally transmit signals as well as those which occupy the gaps between bands also occupy this category of MMs. Figure 3.14 depicts (a) a schematic and fabricated model of a multiband THz antenna [90], and (b) a spring mass model used for band-gap prediction [91].

These materials range in composition, being created using standard printed circuit board (PCB) materials such as silicon, copper, and polytetrafluoroethylene as a substrate. Other materials are printed using additive manufacturing from thermoplastic material as found in the paper surveyed.

3.7 Meta-antenna

Communication between machines takes place over radio waves, which are classified from 3 to 300 GHz. These radio waves are then further divided and classified into various communication bands [92–95]. When sending communications from one device that signal will be received in one form and altered and transmitted in different forms up until it reaches its destination. Research regarding meta-antennas is their ability to receive and transmit signals of multiple bands simultaneously [96–99]. Additionally, they are developed to transcend radio spectrum and developed to communicate in the infrared range using THz frequencies [100–103]. Figure 3.15 from [100] provides an example of a THz antenna, where the oscillating frequencies in the x and y directions (a, c) across the two resonators cause destructive interference in the scattering field narrowing the window in which resonance can occur across the multipole decomposition (b). The power distribution in the y direction produced a more complex current distribution resulting in stronger coupling of the scattered power than the x direction (d). This scattering pattern allows

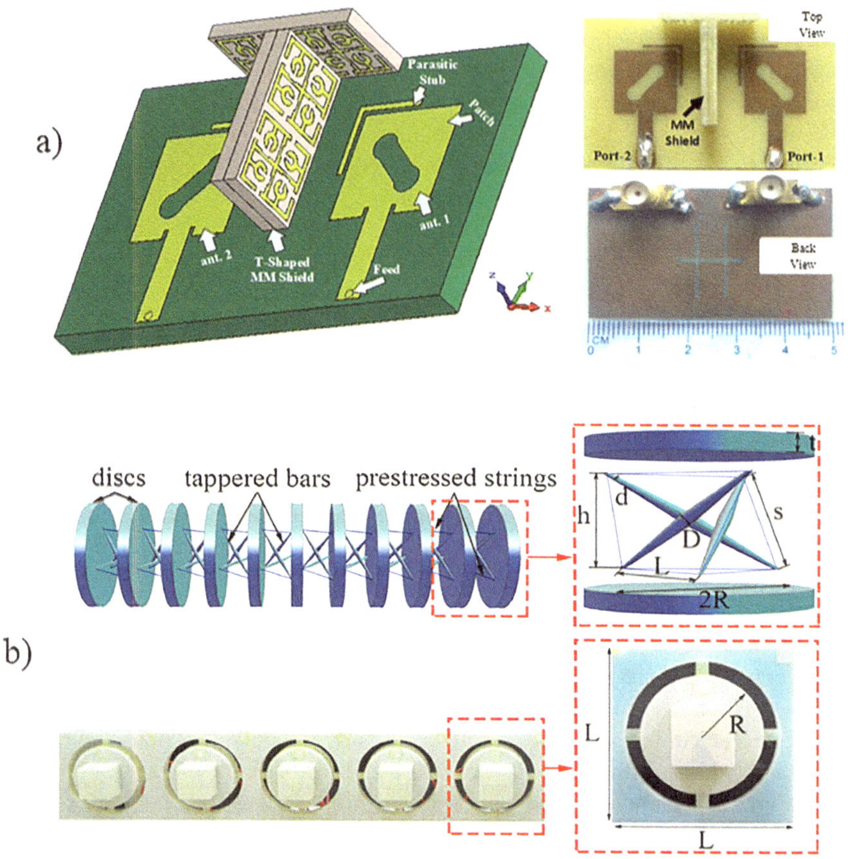

Fig. 3.14 Signal driven transmission based materials: (**a**) Multiband antenna adapted from [90]. (**b**) Band-gap structure adapted from [91]

for the interaction with higher frequency waves as well as the interpretation and decoding of signals within these waves useful for the development of biomedical sensors detecting the presence of biomolecules in cell cultures without the use of higher band X-ray microscopy.

Aside from THz frequencies, meta-antennas can be produced for multiband applications, useful for receiving multiple signals at a time. Figure 3.16. provides an example of schematic representation of a multiband meta-antenna (a) with the simulated results from the proposed antenna (b), a fabricated sample from polytetrafluoroethylene and milled copper (c), along with its experimental measured results showing the same band delineations as the simulated results (d) [97]. This design showed a wide range of interaction between common spectrums used in communication, being able to pick up signals both local internet, radio, and satellite signal simultaneously. However, this does run into a slight problem of size. Now the size

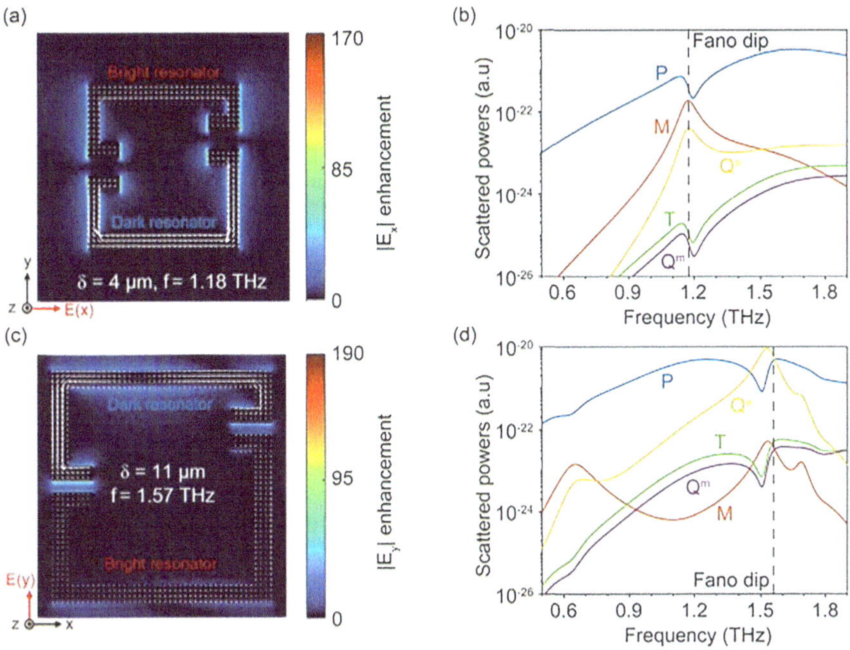

Fig. 3.15 Distribution of field enhancement with induced current density for antenna polarized in *x* (**a**, **b**) and polarized in *y* (**c**, **d**) adapted from [102]

of this antenna, while only 3 cm in length, is still much larger than the common antenna used in phones, routers, and many other portable electronic devices. Further work would yield smaller antenna with comparable capabilities, but their constructions and material considerations would need to be more robust than what was presented in this work.

Meta-antennas are powerful and versatile MMs used in applications such as imaging, multiband data communication, and sensors, among others [104–107]. From the papers surveyed, these antennas are primarily composed of electromagnetic MMs, and acoustic MMS. These materials are often printed on polytetrafluoroethylene boards with copper traces; however, many will combine graphene, silicon, and vanadium dioxide to improve the quality of the frequency transmitted and received by the antenna [108–111]. The geometry used in these antennae can vary widely; however, they follow similar principles in defining boundary conditions to determine the desired frequency range expected for the MM, by ensuring that the antenna strips are thin enough to capture the wavelength of the incident signal [112–114]. The Drude model is a classical model used in modern physics to describe the transport properties of electrons. In this way the permittivity of a material can be found by examining the charge and frequency of the electrons passing through the material. This behavior can be actively described through this model following [115]:

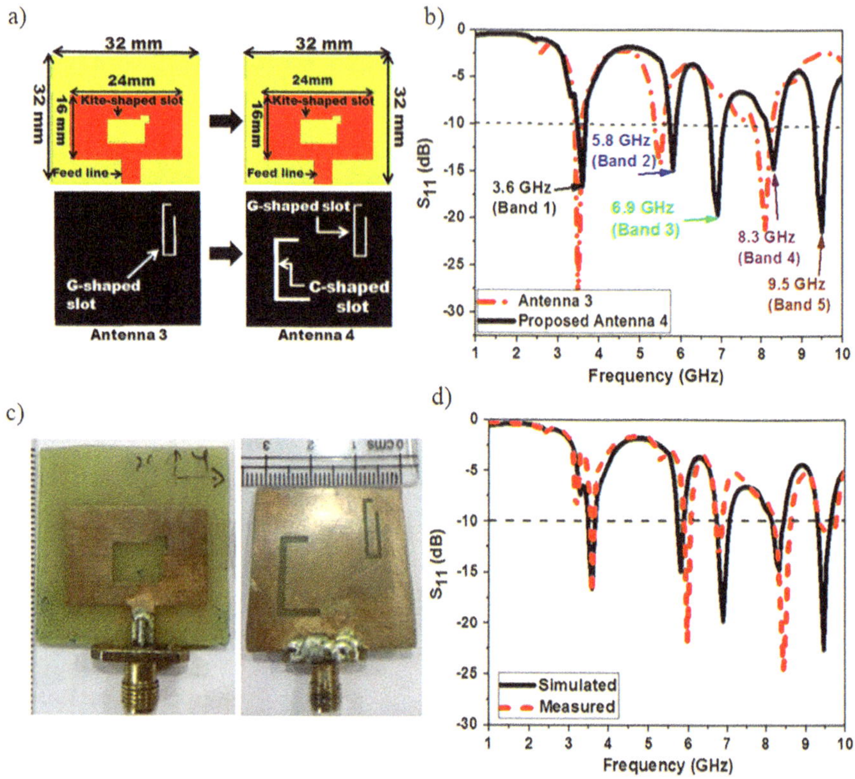

Fig. 3.16 Multiband meta-antenna schematic (a) with proposed structure and simulated results with clear delineation between bands (b) and the physical antenna produced for validation (c) with experimental testing result with multiple clear boundaries between the frequency ranges that can be observed from the experimental testing (d) adapted from [99]

$$\in(\omega) = 1 - \frac{n_e q^2}{\in_0 m\omega^2} \tag{3.5}$$

Several methods to adapt this model have been created, using various material combinations, such as graphene polymerized palm fiber substrates to control the permittivity of the material [116, 117]. Other methods change the geometry of the antenna to create polarization effects on the waves, allowing them to capture ultra-wide band signals as opposed to a narrow range [118].

To assess the design of a novel meta-antenna for THZ applications, a study was launched to embed waveguides into the substrate of the metamaterial [119]. Their material uses two simple antenna shapes overlayed on each other with a sheet of aluminum sandwiched between two layers of silicon. The bottom antenna shape is a simple cross where the top is four circles spaced equidistant about the junction point of the cross below it with "+" shapes inside each circle. The material was fabricated using complementary metal-oxide-semiconductor technology. The permeability and permittivity were measured using the retrieval extraction algorithm and showed the structure to have negative permittivity for the performance range of 0.285 and 0.325 THz.

3.8 Asymmetric Transmission Materials

Asymmetric transmission MMs are a class of MMs that transmit frequencies in one direction and not in another. This means that these materials are either universal receivers or transmitters but cannot perform both actions [120–122]. In the physical sense, these materials can be thought of as diodes for waves rather than electrons. Any of the four major categories of MMs can produce asymmetric transmission materials; this can be performed by a variety of methods including physical deformation of the material, layers of phononic crystals, conversely layered resonators, among many other configurations [123–125]. Often, 3D printing is the best suited methods to produce these materials after numerical analysis because of their complex structural geometry [126, 127]. Among these materials, the fabrication and study of phononic crystals is of great interest. Phononic crystals are periodically arranged scattering materials, wherein their arrangement allows for the control over the band gap for which the material operates to guide the propagation of waves [128, 129]. This type of effect is demonstrated in Fig. 3.17 as the spring mass system demonstrated creates a phononic diode. When excited in the primary direction, the signal is permitted to travel through the material completely (Fig. 3.17a); however, when the signal is reversed the material does not permit the signal to travel through it (Fig. 3.17b) [130].

The literature suggests that the most common type of MM that makes use of asymmetrically transmitting materials would be acoustic MMs. One unique approach validated an asymmetric acoustic MM to determine its sound transmission loss across a wide spectrum of frequencies [131]. The novelty of this material is its expansion of the labyrinthine channels seen in cloaking acoustic MMs. These channels can be used in a linear pattern which allows sound to be transmitted in one direction through the material but becomes trapped and unable to propagate in other directions due to the winding channels within the material.

These channels form new high STL regions and can be explained as the destructive interference among periodic unit cells. The radiation intensities in this material are functions of the unit cell distance and size, as well as their corresponding radiator periods. The resulting radiated sound power interference between unit cells was modeled as [131]:

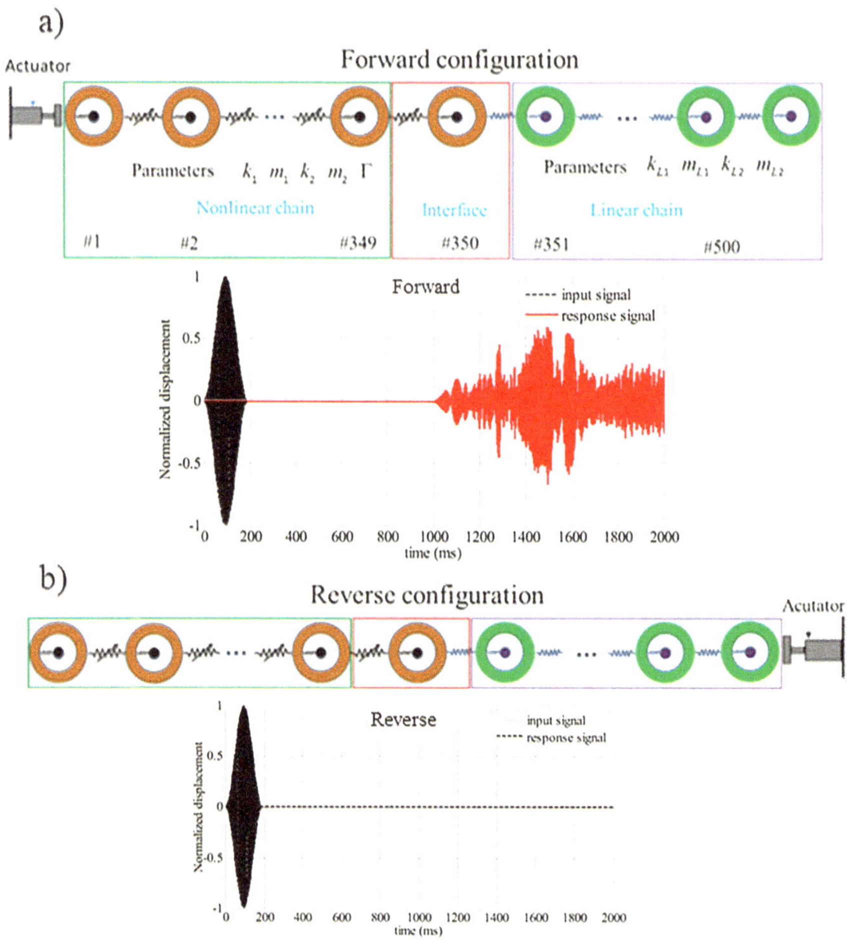

Fig. 3.17 Diagram and demonstration of asymmetric transmission material. (**a**) Illustrates forward behavior allowing for the transmission of signal. (**b**) Illustrates reverse configuration trapping signal preventing it from propagating adapted from [92]

$$W(\alpha,\omega) = \int_{-\frac{\pi}{2}}^{\frac{\pi}{2}} N^2 D(\theta,\alpha,\omega)^2 \, d\theta = \int_{-\frac{\pi}{2}}^{\frac{\pi}{2}} \left| \frac{\sin\left(\frac{Nkb}{2}(\sin\theta + \sin\alpha)\right)}{\sin\left(\frac{kb}{2}(\sin\theta + \sin\alpha)\right)} \right|^2 d\theta \tag{3.6}$$

Using this equation, the stop band for the material was found to be 780 Hz for peak sound transmission loss. Similar studies have been done with auxetic acoustic MMs and gradient-based acoustic MMs to create a phononic crystal effect within the material such that deformation and scattering in the material creates the same asymmetric effect [132, 133].

3.9 Thermal Emissive Materials

Thermal emission metamaterials are developed to control the emission of thermal radiation. Many of the thermal MMs are developed for thermophotovoltaic applications utilizing the conversion of energy from the thermal energy radiating from the thermal MM into electric energy using a photovoltaic cell [134]. These materials can also be used as sensors for gas detection. These materials are designed to work by emitting wavelengths of radiation in each range of temperatures. This emission is then converted to electricity via photovoltaic cells absorbing the radiation [135, 136]. Figure 3.18 outlines the typical process for which a thermal emissions metamaterial will interact with a system [135]. The emitter will generate heat from a

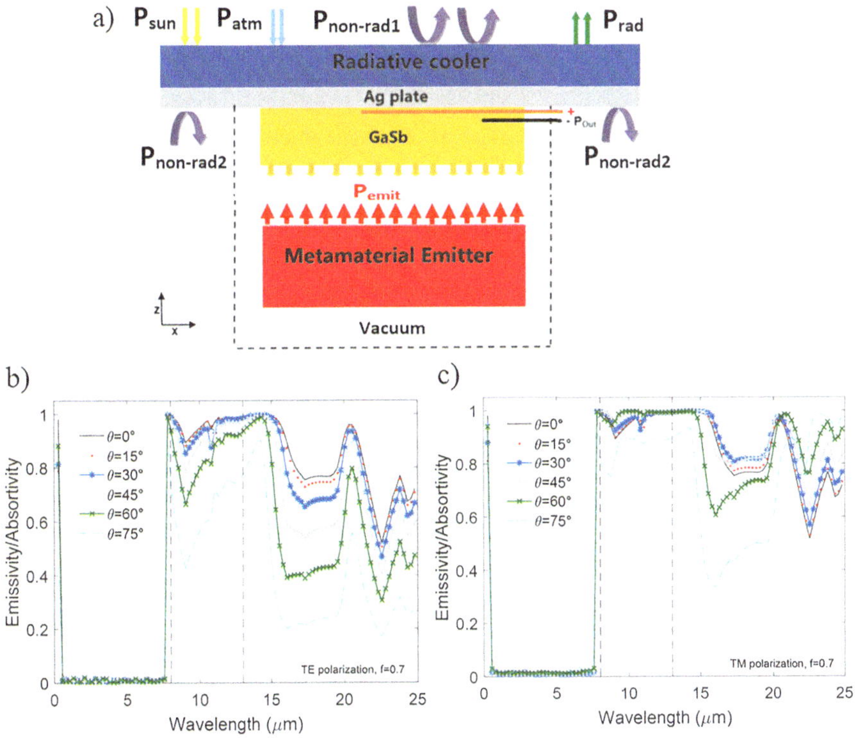

Fig. 3.18 (**a**) Diagram for thermal emitter photovoltaic conversion process with (**b**) thermal electric absorptivity at multiple incident angles for a range of wavelengths from 0 to 25 μm, and (**c**) thermal magnetic absorptivity at multiple incident angles for the same range of wavelengths adapted from [135]

source and radiate infrared waves onto a photovoltaic absorber which then converts that energy into electrical power. Figure 3.18b, c illustrate the absorptivity from the emitter at different angles for a range of wavelengths. It can be observed that shallow angles, 0°, 15°, and 30° produce the best absorptivity regardless of wavelength, but all the incident angles produce the most optimal absorbance between 10 and 15 μm.

As an example, a graphene-based thermal emitter using ring resonators of alternating sizes of a 2 × 2 unit cell grid was developed [137]. The two-dimensional graphene sheet was modeled in terms of its conductivity, carrier scattering, and chemical potential in the graphene with the first term of the polynomial describing the intraband contribution, and the second term describing the interband carrier transitions. These wavelengths were shown to correspond with the resonant frequencies of the ring resonators [137]:

$$\sigma_d = \frac{ie^2 K_b U}{\pi\hbar^2 (\omega + i2\Gamma)} \left(\frac{\mu_c}{K_b U} + 2\ln\left(e^{-\frac{\mu_c}{K_b U}} + 1 \right) \right) + \frac{ie^2}{4\pi\hbar} \ln\left(\frac{2\#\mu_c\# - (\omega + i2\Gamma)\hbar}{2|\mu_c| + (\omega + i2\Gamma)\hbar} \right) \tag{3.7}$$

$$f_o = \frac{nc}{2\pi r\sqrt{\varepsilon_{eff}}} \tag{3.8}$$

The peek emissions of the material were located at arbitrary frequencies through the geometric parameters of the material. In an experiment evaluating the measured reflectance of the thermal emitter, the reflectance was much higher than simulated results across two distinct frequency bands.

3.10 Summary

The dynamic capabilities of energy manipulating metamaterials, a transformative class of materials poised to revolutionize the way energy is managed within various engineering applications, has been explored. By tailoring the composition and geometries of these materials, the ability to absorb specific types of energy with exceptional efficiency has been discovered. This precise control over energy interactions enables the development of innovative applications across multiple fields.

One of the key applications of these materials is the creation of meta-antennas used to transmit and receive many signals simultaneously from multiple frequency bands which can significantly impact telecommunications. Additionally, energy manipulating metamaterials are critical in developing impact-resistant components that offer enhanced safety and durability for automotive and aerospace industries. Thermal management systems designed with these materials can lead to more sustainable and effective control of heat in electronics, machinery, and even large-scale industrial processes.

Energy manipulating metamaterials not only hold the potential to initiate a revolution in engineering applications but also offer a path toward more functional and adaptive solutions. As these materials continue to evolve, they are expected to play a pivotal role in enhancing the performance and efficiency of future technologies, contributing to more innovative, sustainable, and effective engineering solutions. The integration of multifunctional energy manipulating materials into new systems is set to usher in a new era of efficiency and functionality. Systems equipped with these tailored materials are expected to perform at higher standards, meeting the rigorous demands of modern applications in fields such as automotive, electronics, robotics, communications, and beyond.

References

1. Wang, L., Qu, S., Fu, H., Zhou, X., Ding, J., Yang, H., Zhao, Q., Song, X., & Lu, Y. (2024). High-precision Cu alloy microlattices with superior energy absorption capacity enabled by nanoprecipitation engineering. *Scripta Materialia, 239*, 115801.
2. Xiao, H., Li, T., Zhang, L., Liao, W.-H., Tan, T., & Yan, Z. (2023). Metamaterial based piezoelectric acoustic energy harvesting: Electromechanical coupled modeling and experimental validation. *Mechanical Systems and Signal Processing, 185*, 109808.
3. Huang, W., & Zhu, Z. (2023). Broadband metamaterial absorbers based on magnetic composites. *Journal of Magnetism and Magnetic Materials, 576*, 170792.
4. Wu, Z., Zhu, B., Wang, R., & Zhang, X. (2022). Design of mechanical metamaterial for energy absorption using a beam with a variable cross-section. *Mechanism and Machine Theory, 176*, 105027.
5. Elsharabasy, A., Bakr, M., & Deen, M. J. (2020). Wide-angle, wide-band, polarization-insensitive metamaterial absorber for thermal energy harvesting. *Scientific Reports, 10*(1), 16215.
6. Hu, J., Tan, A. T. L., Chen, H., & Hu, X. (2022). Superior compressive properties of 3D printed plate lattice mechanical metamaterials. *International Journal of Mechanical Sciences, 231*, 107586.
7. Wu, Y., Liu, X., Wang, Z., & Ma, G. (2023). Effective broadband electromagnetic wave absorption performance of nano-Fe3O4 magnetic fluid-modified resin geopolymer by metamaterial with a novel patterned structure. *Journal of Magnetism and Magnetic Materials, 580*, 170939.
8. Qi, B., Shou, H., Zhang, J., Chen, W., Feng, J., Niu, T., & Mei, Z. (2023). A near-perfect metamaterial selective absorber for high-efficiency solar photothermal conversion. *International Journal of Thermal Sciences, 194*, 108580.
9. Rui Liu, C., Hui Wu, J., Yang, Z., & Ma, F. (2020). Ultra-broadband acoustic absorption of a thin microperforated panel metamaterial with multi-order resonance. *Composite Structures, 246*, 112366.
10. Qin, J., Liang, K., Deng, X., Li, J., He, C., Liang, X., Jin, L., Zhou, P., & Yu, L. (2024). The chiroptical properties of plasmonic dimer metamaterials based on the extended Born–Kuhn model. *Optics Communications, 553*, 130116.
11. Liang, D., & Wang, C. (2024). Global energy-preserving local mesh-refined S-FDTD schemes for two dimensional Maxwell's equations in Drude metamaterials. *Computer Physics Communications, 294*, 108936.
12. Huang, S., Chen, Y., Yu, C., Chen, S., Zhou, Z., Liang, J., & Dai, W. (2023). Optimized metamaterial solar absorber with ultra-wideband, polarization-independent and large incident angle-insensitive. *Chinese Journal of Physics, 89*, 740.

13. Yang, W., Lu, J., Zhuang, W., Qi, J., Wang, C., Wang, H., Su, G., Xiong, K., Mao, Y., Gong, X., Yuan, J., Wang, D., Sui, J., Zhang, G., Wan, Y., & Yang, P. (2023). PtS2 metamaterials: Fabrication and physical properties. *Applied Surface Science, 614*, 156277.
14. Liu, S., Zhang, L., Bai, G. D., & Cui, T. J. (2019). Flexible controls of broadband electromagnetic wavefronts with a mechanically programmable metamaterial. *Scientific Reports, 9*(1), 1809.
15. Zhou, Q., Liu, H., Gu, Y., Duan, W., Liu, X., Ye, F., Fan, X., & Du, L. (2024). 3D printed PyC/Al2O3 ceramic metamaterials with different micro-channels for tunable microwave absorption. *Journal of the European Ceramic Society, 44*(1), 270–276.
16. Pan, M., Weng, J., Zhou, B., Su, J., Tang, H., Fan, B., Jiang, L., Lin, L., & Qiu, Y. (2023). Terahertz bandwidth absorption metamaterial based on composite structure of single layer graphene. *Diamond and Related Materials, 140*, 110535.
17. Zhang, B., Li, Z., Hu, Z., Zhang, J., & Wang, J. (2022). Analysis of a bidirectional metamaterial perfect absorber with band-switchability for multifunctional optical applications. *Results in Physics, 34*, 105313.
18. Singh, G., Bhardwaj, A., Srivastava, K. V., Ramkumar, J., & Ramakrishna, S. A. (2021). Perforated lightweight microwave metamaterial broadband absorber with discontinuous ground plane. *Applied Physics A, 127*(11), 858.
19. Zhu, J., Wu, C., & Ren, Y. (2021). Broadband terahertz metamaterial absorber based on graphene resonators with perfect absorption. *Results in Physics, 26*, 104466.
20. Ullah, N., Shabiul Islam, M., Hoque, A., Yong, W. H., Soliman, M. S., & Islam, M. T. (2024). A compact-sized four-band metamaterial-based perfect absorber for electromagnetic energy harvesting applications. *Optics & Laser Technology, 168*, 109836.
21. Al-Badri, K. S. L. (2021). Microwave metamaterial for broad-band perfect absorber applications. *Materials Today: Proceedings, 42*, 2835–2839.
22. Petronijevic, E., & Sibilia, C. (2020). Thin films of phase change materials for light control of metamaterials in the optical and infrared spectral domain. *Optical and Quantum Electronics, 52*(2), 110.
23. Gandhi, C., Babu, P. R., & Senthilnathan, K. (2021). Ultra-thin polarization independent broadband terahertz metamaterial absorber. *Frontiers of Optoelectronics, 14*(3), 288–297.
24. Moniruzzaman, M., Islam, M. T., Hossain, I., Soliman, M. S., Samsuzzaman, M., & Almalki, S. H. A. (2021). Symmetric resonator based tunable epsilon negative near zero index metamaterial with high effective medium ratio for multiband wireless applications. *Scientific Reports, 11*(1), 21842.
25. Chen, F., Gupta, N., Behera, R. K., & Rohatgi, P. K. (2018). Graphene-reinforced aluminum matrix composites: A review of synthesis methods and properties. *JOM, 70*(6), 837–845.
26. Poveda, R.L. and N. Gupta, Electrical properties of carbon nanofiber reinforced multiscale polymer composites. Materials & Design (1980–2015), 2014. 56: p. 416–422.
27. Singh, A. K., Shishkin, A., Koppel, T., & Gupta, N. (2018). A review of porous lightweight composite materials for electromagnetic interference shielding. *Composites Part B: Engineering, 149*, 188–197.
28. Borzooei, S., Rezagholizadeh, E., & Biabanifard, M. (2020). Graphene disks for frequency control of terahertz waves in broadband applications. *Journal of Computational Electronics, 19*(2), 759–772.
29. Liu, H., Zhang, Y., Liu, X., Duan, W., Li, M., Zhou, Q., Li, S., Wang, G., & Han, G. (2022). Additive manufacturing of nanocellulose/polyborosilazane derived CNFs-SiBCN ceramic metamaterials for ultra-broadband electromagnetic absorption. *Chemical Engineering Journal, 433*, 133743.
30. Zhou, Q., Shi, T., Xue, B., Gu, S., Ren, W., Ye, F., Fan, X., & Du, L. (2023). Multi-scale integrated design and fabrication of ultra-broadband electromagnetic absorption utilizing multi-walled carbon nanotubes-based hierarchical metamaterial. *Composites Science and Technology, 232*, 109877.

31. Wen, J., Sun, W., Liang, B., He, C., Xiong, K., Wu, Z., Zhang, H., Yu, H., Wang, Q., Pan, Y., Zhang, Y., & Liu, Z. (2023). Dynamically switchable broadband–narrowband terahertz metamaterial absorber based on vanadium dioxide and multilayered structure. *Optics Communications, 545*, 129710.
32. Zhong, M. (2023). Measurement and verification of a multi-band terahertz metamaterial absorber based on multiple coupling effects. *Infrared Physics & Technology, 128*, 104506.
33. Ullah, N., Islam, M. S., Hoque, A., Yong, W. H., Alrashdi, A. M., Soliman, M. S., & Islam, M. T. (2023). An efficient, compact, wide-angle, wide-band, and polarization-insensitive metamaterial electromagnetic energy harvester. *Alexandria Engineering Journal, 82*, 377–388.
34. Shi, Z., Song, L., & Zhang, T. (2019). Optical and electrical characterization of pure PMMA for terahertz wide-band metamaterial absorbers. *Journal of Infrared, Millimeter, and Terahertz Waves, 40*(1), 80–91.
35. Sun, X., Wang, Z., Wang, S., Ning, Y., Yang, M., Yang, S., Zhou, L., He, Q., & Li, Y. (2021). Ultrabroad-band and low-frequency microwave absorption based on activated waxberry metamaterial. *Chemical Engineering Journal, 422*, 130142.
36. Yang, L., Wang, H., Ren, X., Song, X., Chen, M., Tong, Y., Ye, Y., Ren, Y., Liu, S., Wang, S., Yin, L., & Yao, J. (2021). Switchable terahertz absorber based on metamaterial structure with photosensitive semiconductor. *Optics Communications, 485*, 126708.
37. Malekshahi, N., Amini, A., & Chaharmahali, I. (2021). A multilayer graphene-based absorber for the sub-THz regime. *Journal of Computational Electronics, 20*(3), 1343–1353.
38. Barimani, S., Amini, A., & Chaharmahali, I. (2021). A four-bias three-layer graphene-based THz absorber. *Journal of Computational Electronics, 20*(3), 1332–1342.
39. Du, L., Shi, T., Zhou, Q., Li, Y., Zhang, H., Dong, S., Deng, L., Zhao, J., Li, X., Wang, J., & Zou, C. (2023). A multi-layer square frustum metamaterial for ultra-broadband electromagnetic absorption based on carbonyl iron powder/carbon fiber composites. *Journal of Alloys and Compounds, 950*, 169917.
40. Yan, L., Wang, Y., Li, W., Yin, S., Huang, W., Yin, Z., Jia, D., & Li, Y. (2023). A bilayer array metamaterial based on silicon carbon foam/FeSiAl for broadband electromagnetic absorption. *Journal of Alloys and Compounds, 954*, 170129.
41. Ma, X., Duan, Y., Huang, L., Lei, H., & Yang, X. (2022). Quasiperiodic metamaterials with broadband absorption: Tailoring electromagnetic wave by Penrose tiling. *Composites Part B: Engineering, 233*, 109659.
42. Zega, V., Silva, P. B., Geers, M. G. D., & Kouznetsova, V. G. (2020). Experimental proof of emergent subharmonic attenuation zones in a nonlinear locally resonant metamaterial. *Scientific Reports, 10*(1), 12041.
43. Wang, S., Zhang, X., Li, F., & Hosseini, S. M. (2022). Sound transmission loss of a novel acoustic metamaterial sandwich panel: Theory and experiment. *Applied Acoustics, 199*, 109035.
44. Fan, J., Zhang, L., Wang, X., Zhang, Z., Wei, S., Song, B., Zhao, A., Xiang, X., Zhu, X., & Shi, Y. (2022). 3D printed ultra-thin acoustic metamaterials with adaptable low-frequency absorption performance. *Chinese Journal of Mechanical Engineering: Additive Manufacturing Frontiers, 1*(3), 100036.
45. de Melo Filho, N. G. R., Claeys, C., Deckers, E., & Desmet, W. (2020). Metamaterial foam core sandwich panel designed to attenuate the mass-spring-mass resonance sound transmission loss dip. *Mechanical Systems and Signal Processing, 139*, 106624.
46. Liu, Q., Liu, X., Zhang, C., & Xin, F. (2021). High-temperature and low-frequency acoustic energy absorption by a novel porous metamaterial structure. *Acta Mechanica Solida Sinica, 34*(6), 872–883.
47. Li, K., Nennig, B., Perrey-Debain, E., & Dauchez, N. (2021). Poroelastic lamellar metamaterial for sound attenuation in a rectangular duct. *Applied Acoustics, 176*, 107862.

48. Zhao, B., Thomsen, H. R., De Ponti, J. M., Riva, E., Van Damme, B., Bergamini, A., Chatzi, E., & Colombi, A. (2022). A graded metamaterial for broadband and high-capability piezoelectric energy harvesting. *Energy Conversion and Management, 269*, 116056.
49. Hu, G., Tang, L., & Cui, X. (2019). On the modelling of membrane-coupled Helmholtz resonator and its application in acoustic metamaterial system. *Mechanical Systems and Signal Processing, 132*, 595–608.
50. Zhong, J., Chai, Z., & Xiang, J. (2023). Double defects in trampoline effect and Helmholtz coupled acoustic metamaterial for broadband piezoelectric energy harvesting. *Wave Motion, 124*, 103248.
51. Liu, Y., Zhao, D., Yan, Z., Sun, W., Guo, P., & Tan, T. (2023). Reprogrammable acoustic metamaterials for multiband energy harvesting. *Engineering Structures, 288*, 116207.
52. Ma, K., Tan, T., Yan, Z., Liu, F., Liao, W.-H., & Zhang, W. (2021). Metamaterial and Helmholtz coupled resonator for high-density acoustic energy harvesting. *Nano Energy, 82*, 105693.
53. Amaral, D. R., Ichchou, M. N., Kołakowski, P., Fossat, P., & Salvia, M. (2023). Lightweight gearbox housing with enhanced vibro-acoustic behavior through the use of locally resonant metamaterials. *Applied Acoustics, 210*, 109435.
54. Nimmagadda, C., & Matlack, K. H. (2019). Thermally tunable band gaps in architected metamaterial structures. *Journal of Sound and Vibration, 439*, 29–42.
55. Gui, Y., Fei, Z., Zhao, S., Zhang, Z., Shao, H., Chen, J., Li, K., Du, H., & Yang, Z. (2023). High-strength and multifunctional honeycomb polyimide aerogel fabricated by a freeze casting-assisted extrusion printing and building block-assembly strategy for sound absorbing metamaterials. *Additive Manufacturing, 77*, 103799.
56. Brooke, D. C., Umnova, O., Leclaire, P., & Dupont, T. (2020). Acoustic metamaterial for low frequency sound absorption in linear and nonlinear regimes. *Journal of Sound and Vibration, 485*, 115585.
57. Gao, C., Halim, D., & Yi, X. (2022). Elastic metamaterial with multiple resonant modes and asymmetric structure design for low-frequency vibration absorption. *Acta Mechanica, 233*(12), 5321–5333.
58. Sheng, P., Fang, X., Dai, L., Yu, D., & Wen, J. (2023). Synthetical vibration reduction of the nonlinear acoustic metamaterial honeycomb sandwich plate. *Mechanical Systems and Signal Processing, 185*, 109774.
59. de Melo Filho, N. G. R., Van Belle, L., Claeys, C., Deckers, E., & Desmet, W. (2019). Dynamic mass based sound transmission loss prediction of vibro-acoustic metamaterial double panels applied to the mass-air-mass resonance. *Journal of Sound and Vibration, 442*, 28–44.
60. Chu, J., Zhou, G., Liang, X., Liang, H., Yang, Z., & Chen, T. (2023). A metamaterial for low-frequency vibration damping. *Materials Today Communications, 36*, 106464.
61. Sheng, P., Fang, X., Wen, J., & Yu, D. (2021). Vibration properties and optimized design of a nonlinear acoustic metamaterial beam. *Journal of Sound and Vibration, 492*, 115739.
62. Basta, E., Ghommem, M., & Emam, S. (2020). Vibration suppression of nonlinear rotating metamaterial beams. *Nonlinear Dynamics, 101*(1), 311–332.
63. Porubov, A. V. (2023). Nonlinear wave localization in an acoustic metamaterial. *Communications in Nonlinear Science and Numerical Simulation, 119*, 107095.
64. Lee, S., Ahn, C. H., & Lee, J. W. (2018). Vibro-acoustic metamaterial for longitudinal vibration suppression in a low frequency range. *International Journal of Mechanical Sciences, 144*, 223–234.
65. Lin, Q., Lin, Q., Wang, Y., & Di, G. (2021). Sound insulation performance of sandwich structure compounded with a resonant acoustic metamaterial. *Composite Structures, 273*, 114312.
66. Jang, J.-Y., & Song, K. (2023). Synergistic acoustic metamaterial for soundproofing: Combining membrane and locally resonant structure. *International Journal of Mechanical Sciences, 256*, 108500.

67. Sun, P., Xu, S., Wang, X., Gu, L., Luo, X., Zhao, C., & Huang, Z. (2023). Sound absorption of space-coiled metamaterials with soft walls. *International Journal of Mechanical Sciences*, 108696.
68. Yang, C., Zhong, T., Li, J., You, S., Yang, S., Zhang, H., & Zheng, Z. (2022). Extended tube acoustic metamaterial: Its modeling and application to a kitchen hood. *Applied Acoustics, 185*, 108398.
69. Jia, X., Yan, M., & Hong, M. (2021). Sound energy enhancement via impedance-matched anisotropic metamaterial. *Materials & Design, 197*, 109254.
70. Jang, J.-Y., Park, C.-S., & Song, K. (2022). Lightweight soundproofing membrane acoustic metamaterial for broadband sound insulation. *Mechanical Systems and Signal Processing, 178*, 109270.
71. Lau, S.-K., Zhu, X.-F., & Lu, Z. (2021). Enhancing the acoustic absorption of vegetation with embedded periodic metamaterials. *Applied Acoustics, 171*, 107576.
72. Zhong, H., Gu, Y., Bao, B., Wang, Q., & Wu, J. (2019). 2D underwater acoustic metamaterials incorporating a combination of particle-filled polyurethane and spiral-based local resonance mechanisms. *Composite Structures, 220*, 1–10.
73. Ciaburro, G., & Iannace, G. (2022). Membrane-type acoustic metamaterial using cork sheets and attached masses based on reused materials. *Applied Acoustics, 189*, 108605.
74. Aladwani, A., Almandeel, A., & Nouh, M. (2019). Fluid-structural coupling in metamaterial plates for vibration and noise mitigation in acoustic cavities. *International Journal of Mechanical Sciences, 152*, 151–166.
75. Zhong, H., Tian, Y., Gao, N., Lu, K., & Wu, J. (2021). Ultra-thin composite underwater honeycomb-type acoustic metamaterial with broadband sound insulation and high hydrostatic pressure resistance. *Composite Structures, 277*, 114603.
76. Wu, J., Zhang, Y., Yang, F., Jiang, F., Xu, X., Tan, Y., & Su, L. (2023). A hybrid architectural metamaterial combing plate lattice and hollow-truss lattice with advanced mechanical performances. *Additive Manufacturing, 76*, 103764.
77. Tan, X., Wang, B., Yao, K., Zhu, S., Chen, S., Xu, P., Wang, L., & Sun, Y. (2019). Novel multi-stable mechanical metamaterials for trapping energy through shear deformation. *International Journal of Mechanical Sciences, 164*, 105168.
78. Chen, S., Liu, X., Hu, J., Wang, B., Li, M., Wang, L., Zou, Y., & Wu, L. (2023). Elastic architected mechanical metamaterials with negative stiffness effect for high energy dissipation and low frequency vibration suppression. *Composites Part B: Engineering, 267*, 111053.
79. Patil, S., Ganzenmüller, G., Gutmann, F., Hoschke, K., & Hiermaier, S. (2023). The Auxetic friction cell: Towards programming strain rate dependency and energy dissipation into mechanical metamaterials. *Materials Today Communications, 36*, 106725.
80. Han, H., Sorokin, V., Tang, L., & Cao, D. (2023). Origami-based tunable mechanical memory metamaterial for vibration attenuation. *Mechanical Systems and Signal Processing, 188*, 110033.
81. Ma, H., Wang, K., Zhao, H., Shi, W., Xue, J., Zhou, Y., Li, Q., Wang, G., & Yan, B. (2022). Energy dissipation and shock isolation using novel metamaterials. *International Journal of Mechanical Sciences, 228*, 107464.
82. Zhao, P., Zhang, K., Qi, L., & Deng, Z. (2022). 3D chiral mechanical metamaterial for tailored band gap and manipulation of vibration isolation. *Mechanical Systems and Signal Processing, 180*, 109430.
83. Xu, R., He, Y., Li, X., Lu, M., & Chen, Y. (2023). Snap-fit mechanical metamaterials. *Applied Materials Today, 30*, 101714.
84. Wei, Y.-L., Yang, Q.-S., Liu, X., & Tao, R. (2022). Multi-bionic mechanical metamaterials: A composite of FCC lattice and bone structures. *International Journal of Mechanical Sciences, 213*, 106857.
85. Tan, X., Chen, S., Zhu, S., Wang, B., Xu, P., Yao, K., & Sun, Y. (2019). Reusable metamaterial via inelastic instability for energy absorption. *International Journal of Mechanical Sciences, 155*, 509–517.

86. Wang, X., & Wu, J. (2021). Uncooled CMOS 2.5THz detector comprising a metamaterial absorber loaded with resistors and a PTAT sensor. *Optical and Quantum Electronics, 53*(5), 217.
87. Bai, J., Pang, Z., Shen, P., Chen, T., Shen, W., Wang, S., & Chang, S. (2021). A terahertz photo-thermoelectric detector based on metamaterial absorber. *Optics Communications, 497*, 127184.
88. Zhao, S., Ma, H., Shao, T., Wang, J., Yang, Z., Meng, Y., Feng, M., Yan, M., Wang, J., & Qu, S. (2021). High temperature metamaterial enhanced electromagnetic absorbing coating prepared with alumina ceramic. *Journal of Alloys and Compounds, 874*, 159822.
89. Xu, C., Qu, S., Pang, Y., Wang, J., Yan, M., Zhang, J., Wang, Z., & Wang, W. (2018). Metamaterial absorber for frequency selective thermal radiation. *Infrared Physics & Technology, 88*, 133–138.
90. Hasan, M. M., Islam, M. T., Alam, T., Alzamil, A., & Soliman, M. S. (2024). Electromagnetic coupling shielding in compact MIMO antenna using symmetric T-shaped metamaterial structure for 5G communications. *Optics & Laser Technology, 169*, 110046.
91. Vo, N. H., Pham, T. M., Hao, H., Bi, K., & Chen, W. (2022). A reinvestigation of the spring-mass model for metamaterial bandgap prediction. *International Journal of Mechanical Sciences, 221*, 107219.
92. Singh, G., Abrol, A., Kumar, S., Kumar Kanaujia, B., & Kumar Pandey, V. (2023). Electromagnetic metamaterial-inspired wideband millimeter-wave antenna for 5G communication. *Materials Today: Proceedings*.
93. De Miguel-Hernández, J., Hoyland, R. J., Sosa-Cabrera, D., Deviaene, S., Fuerte-Rodríguez, P. A., González-Carretero, E. D., & Vega-Moreno, A. (2019). Manufacturing of 3D-metallic electromagnetic metamaterials for feedhorns used in radioastronomy and satellite communications. *Mechanics of Materials, 139*, 103195.
94. Tamim, A. M., Faruque, M. R. I., Khandaker, M. U., Islam, M. T., & Bradley, D. A. (2021). Electromagnetic radiation reduction using novel metamaterial for cellular applications. *Radiation Physics and Chemistry, 178*, 108976.
95. Ramachandran, T., Faruque, M. R. I., & Ahamed, E. (2019). Composite circular split ring resonator (CSRR)-based left-handed metamaterial for C- and Ku-band application. *Results in Physics, 14*, 102435.
96. Hossen, S., Alqahtani, A., Hossain, I., Islam, M. T., Moniruzzaman, M., & Samsuzzaman, M. (2023). Gap coupled symmetric square split ring high EMR resonator-based metamaterial for S-, C-, and X-bands wireless applications: Simulation and experiment. *Optical Materials, 145*, 114389.
97. Ali, T., Prasad, K. D., & Biradar, R. C. (2018). A miniaturized slotted multiband antenna for wireless applications. *Journal of Computational Electronics, 17*(3), 1056–1070.
98. Ali, W. A. E., Mohamed, H. A., Ibrahim, A. A., & Hamdalla, M. Z. M. (2019). Gain improvement of tunable band-notched UWB antenna using metamaterial lens for high speed wireless communications. *Microsystem Technologies, 25*(11), 4111–4117.
99. Katare, K. K., Biswas, A., & Akhtar, M. J. (2019). Near-field phase modulation using a semicircular radially gradient metasurface for beam steering of an RF antenna. *Journal of Computational Electronics, 18*(2), 671–679.
100. Ryu, H., Kang, J.-H., & Lee, S.-H. (2023). Terahertz metamaterials for biomolecule sensing based on symmetry-broken unit resonators. *Results in Physics, 54*, 107049.
101. Malhotra, I., Jha, K. R., & Singh, G. (2018). Design of highly directive terahertz photoconductive dipole antenna using frequency-selective surface for sensing and imaging applications. *Journal of Computational Electronics, 17*(4), 1721–1740.
102. He, X., Wang, Y., Tao, M., Yu, X., Pei, Z., Wang, B., Yang, Y., Jiang, J., & Geng, Z. (2019). Dynamical switching of electromagnetically induced reflectance in complementary terahertz metamaterials. *Optics Communications, 448*, 98–103.
103. Geetharamani, G., & Aathmanesan, T. (2019). Metamaterial inspired THz antenna for breast cancer detection. *SN Applied Sciences, 1*(6), 595.

104. Anand, S., & Prashalee, P. (2021). High gain compact multiband cavity-backed SIW and metamaterial unit cells with CPW feed antenna for S, and Ku band applications. *Wireless Personal Communications, 118*(2), 1621–1634.
105. Arumona, A. E., Garhal, A., Punthawanunt, S., Ray, K., Youplao, P., & Yupapin, P. (2020). Micro-metamaterial antenna characteristics using microring embedded silver bars. *Microsystem Technologies, 26*(12), 3927–3933.
106. Patil, K. S., & Rufus, E. (2022). Design of bio-implantable antenna using metamaterial substrate. *Wireless Personal Communications, 124*(2), 1443–1455.
107. Vasu Babu, K., Das, S., Varshney, G., Sree, G. N. J., & Madhav, B. T. P. (2022). A micro-scaled graphene-based tree-shaped wideband printed MIMO antenna for terahertz applications. *Journal of Computational Electronics, 21*(1), 289–303.
108. Alizadeh, A., Nazeri, M., & Sajedi Bidgoli, A. (2020). Enhancement of the frequency peak of terahertz photoconductive antennas using metamaterial (MTM) superstrate structures. *Journal of Computational Electronics, 19*(1), 451–456.
109. Smolyaninov, I. I., & Smolyaninova, V. N. (2020). Hybrid acousto-electromagnetic metamaterial superconductors. *Physica C: Superconductivity and Its Applications, 577*, 1353730.
110. Biabanifard, S. (2021). A graphene-based dual-band THz absorber design exploiting the impedance-matching concept. *Journal of Computational Electronics, 20*(1), 38–48.
111. Xu, J., Cao, J., Guo, M., Yang, S., Yao, H., Lei, M., Hao, Y., & Bi, K. (2021). Metamaterial mechanical antenna for very low frequency wireless communication. *Advanced Composites and Hybrid Materials, 4*(3), 761–767.
112. Fei, S., Guo, S., Li, B., Liu, Y., & He, S. (2019). An acoustic metamaterial lens for acoustic point-to-point communication in air. *Acoustical Physics, 65*(1), 1–6.
113. Zhang, X., & Lin, Y.-S. (2019). Actively electromagnetic modulation of IHI-shaped terahertz metamaterial with high-efficiency switching characteristic. *Results in Physics, 15*, 102532.
114. Torabi, Y., & Omidi, R. (2018). Novel metamaterial compact planar MIMO antenna systems with improved isolation for WLAN application. *Wireless Personal Communications, 102*(1), 399–410.
115. Garhwal, A., Bunruangses, M., Arumona, A. E., Youplao, P., Ray, K., Suwandee, S., & Yupapin, P. (2021). Integrating metamaterial antenna node and LiFi for privacy preserving intelligent COVID-19 hospital patient management. *Cognitive Computation, 16*, 1623.
116. Elwi, T. A. (2020). A further realization of a flexible metamaterial-based antenna on nickel oxide polymerized palm fiber substrates for RF energy harvesting. *Wireless Personal Communications, 115*(2), 1623–1634.
117. Esfandiyari, M., Lalbakhsh, A., Jarchi, S., Ghaffari-Miab, M., Mahtaj, H. N., & Simorangkir, R. B. V. B. (2022). Tunable terahertz filter/antenna-sensor using graphene-based metamaterials. *Materials & Design, 220*, 110855.
118. Li, X., Xu, Y., Wang, H., Zhang, Y., & Lv, G. (2018). Low cross-polarization antipodal tapered slot antenna with gain bandwidth enhancement for UWB application. *Journal of Computational Electronics, 17*(1), 442–451.
119. Alibakhshikenari, M., Virdee, B. S., Althuwayb, A. A., Aïssa, S., See, C. H., Abd-Alhameed, R. A., Falcone, F., & Limiti, E. (2021). Study on on-chip antenna design based on metamaterial-inspired and substrate-integrated waveguide properties for millimetre-wave and THz integrated-circuit applications. *Journal of Infrared, Millimeter, and Terahertz Waves, 42*(1), 17–28.
120. Wu, R.-T., Liu, T.-W., Jahanshahi, M. R., & Semperlotti, F. (2021). Design of one-dimensional acoustic metamaterials using machine learning and cell concatenation. *Structural and Multidisciplinary Optimization, 63*(5), 2399–2423.
121. Jalšić, M., Alujević, N., Garma, T., Ćatipović, I., Jokić, M., & Wolf, H. (2023). An active metamaterial cell concept for nonreciprocal vibroacoustic transmission. *Mechanical Systems and Signal Processing, 186*, 109829.

122. Sampaio, L. Y. M., Rodrigues, G. K., Mosquera-Sánchez, J. A., De Marqui, C., & de Oliveira, L. P. R. (2021). Membrane smart metamaterials for unidirectional wave propagation problems. *Journal of Sound and Vibration, 512*, 116374.
123. Guo, X.-F., & Ma, L. (2020). One-way transmission in topological mechanical metamaterials based on self-locking. *International Journal of Mechanical Sciences, 175*, 105555.
124. Li, Z., Wang, Y., & Wang, Y. (2022). Tunable three-dimensional nonreciprocal transmission in a layered nonlinear elastic wave metamaterial by initial stresses. *Applied Mathematics and Mechanics, 43*(2), 167–184.
125. Mirzamohammadi, F., Nourinia, J., Ghobadi, C., & Naderali, R. (2019). A dual-wideband bi-layered chiral metamaterial to develop cross-polarization conversion and asymmetric transmission functionalities for the linearly polarized electromagnetic waves. *AEU—International Journal of Electronics and Communications, 111*, 152916.
126. Kulkarni, P. P., & Manimala, J. M. (2019). Realizing passive direction-bias for mechanical wave propagation using a nonlinear metamaterial. *Acta Mechanica, 230*(7), 2521–2537.
127. Chen, T., Li, W., & Yu, D. (2021). A tunable gradient acoustic metamaterial for acoustic sensing. *Extreme Mechanics Letters, 49*, 101481.
128. Sridhar, A., Kouznetsova, V. G., & Geers, M. G. D. (2020). Frequency domain boundary value problem analyses of acoustic metamaterials described by an emergent generalized continuum. *Computational Mechanics, 65*(3), 789–805.
129. Miao, Z.-H., & Wang, Y.-Z. (2022). Nonreciprocal transmission of non-collinear mixing wave in nonlinear elastic wave metamaterial. *Journal of Elasticity, 150*(2), 341–366.
130. Zhou, W. J., Li, X. P., Wang, Y. S., Chen, W. Q., & Huang, G. L. (2018). Spectro-spatial analysis of wave packet propagation in nonlinear acoustic metamaterials. *Journal of Sound and Vibration, 413*, 250–269.
131. Yu, X., Lu, Z., Liu, T., Cheng, L., Zhu, J., & Cui, F. (2019). Sound transmission through a periodic acoustic metamaterial grating. *Journal of Sound and Vibration, 449*, 140–156.
132. Cao, W.-K., Wu, L.-T., Zhang, C., Ke, J.-C., Cheng, Q., Cui, T.-J., & Jing, Y. (2019). Asymmetric transmission of acoustic waves in a waveguide via gradient index metamaterials. *Science Bulletin, 64*(12), 808–813.
133. Tao, Z., Ren, X., Zhao, A. G., Sun, L., Zhang, Y., Jiang, W., Han, D., Zhang, X. Y., & Xie, Y. M. (2022). A novel auxetic acoustic metamaterial plate with tunable bandgap. *International Journal of Mechanical Sciences, 226*, 107414.
134. Zhang, Y., Li, K., Yang, X., Cao, S., Pang, H., Cai, Q., Ye, Q., & Wu, X. (2022). Thermal degradation of tungsten nanowire-based hyperbolic metamaterial emitters for near-field thermophotovoltaic applications. *International Journal of Thermophysics, 43*(2), 16.
135. Zahir, M., Drevillon, J., & Ezzahri, Y. (2023). Improved thermophotovoltaic system performance with a novel metamaterial thermal emitter combined with a passive radiative cooling approach. *Solar Energy Materials and Solar Cells, 263*, 112587.
136. Gupta, N.K., H. Wanare, and S.A. Ramakrishna. *Information Encryption in Thermal Metamaterials by Emissivity Engineering*. in 2019 Workshop on Recent Advances in Photonics (WRAP). 2019.
137. Shi, C., Mahlmeister, N. H., Luxmoore, I. J., & Nash, G. R. (2018). Metamaterial-based graphene thermal emitter. *Nano Research, 11*(7), 3567–3573.

Chapter 4
Cloaking Properties of Metamaterials

List of Symbols

α	Angle of incidence
C	Thermal capacity
D_e	Diameter
∈	Permittivity
L	Length
n	Mode number
τ	Pitch
r	Radius
S	Scattering parameter
T	Temperature
z	Impedance
γ	Resonant amplitudes in dark modes
δ	Resonant detuning frequency
κ_c	Coupling coefficient
κ	Thermal conductivity tensor
μ	Permeability
φ	Refractive index
χ	Susceptibility
ω	Angular frequency
k	Stiffness
m_{eff}	Effective mass

4.1 Introduction

Cloaking metamaterials are revolutionary in material science bringing what was often thought to be science fiction into reality by creating these materials which bend waves around them, keeping a central object unaffected by the incident waves.

N. Gupta, C. Beckwith, *Architected Metamaterials: Design Principles and Properties*, https://doi.org/10.1007/978-3-031-98330-6_4

Similar phenomena are used in each of the four major categories of materials though each accomplishes this in unusual ways to produce the desired effect [1–3]. Electromagnetic MM cloaks utilize double negative material behavior such that the permeability and permittivity of the material do not allow electric and/or magnetic fields to pass through the material [4–6]. This is often accomplished through the use of a split ring resonator (SRR) which at certain frequencies causes electromagnetic signals to pass around the resonator like a barrier rather than pass through [7, 8]. Acoustic MMs have multiple approaches to cloaking, some act as a lens to create a waveguide for acoustic waves, or they will take a more geometry based mechanical approach through a labyrinth-like lattice which the waves travel through [9, 10]. Thermal MM cloaks work similar to electromagnetic MM cloaks in that the material does not permit conductive, convective, or radiative waves of heat to pass through the material; in many cases this is done by equalizing thermal and electrical conductivity to create a scattering effect across the material causing incident waves to pass around it [11]. Lastly, mechanical MM cloaks tend to work through manipulation of the stiffness of the material, allowing it to deform in response to forces around a particular area [12, 13]. Mechanical cloaking makes up the smallest portion of the papers surveyed, where the focus surrounding mechanical MMs revolve around the creation of energy absorbing, elastic, or auxetic material where any cloaking effect was secondary to the design of the material.

There are two main types of phenomena that MMs use to create the cloaking effect. Transformation optics is a theoretical framework that manipulates stimuli from MMs with tailored geometries and properties inspired by the theory of general relativity. This approach allows for the control of the paths of the stimuli, enabling the creation of cloaks and super lenses defying the limits of traditional diffraction theory [14–16]. Metamaterials play a crucial role here as they can be designed with specific permeability and permittivity targets to bend light, sound, and heat like how gravity bends spacetime. Electromagnetically induced transparency (EIT), in contrast, is a quantum optical phenomenon that uses light to induce quantum interference making an opaque medium transparent to specific wavelengths of light [17–22]. These methods both work at different scales, EIT working on the microscopic scale, and transformation optics working at the macroscopic scale.

Figure 4.1 provides five examples of cloaking metamaterials. Figure 4.1a provides a visualization of the wave interaction surrounding an acoustic MM where the incident wave is directed around the central object returning to its original propagation path [23]. Figure 4.1b provides an example for the acoustic MM construction, using a labyrinthine construction of corridors to channel the flow of the waves in a desired direction [10]. Figure 4.2a demonstrates a simulation of an electromagnetic MM cloak where the "+-" shaped unit cells on the printed circuit board array with and without the cloak activated at 2.6 gigahertz (GHz) [24]. Figure 4.2b contains a schematic distribution of unit cells for a mechanical cloak composed of various resonators of various distribution cut out from an aluminum sheet for testing against

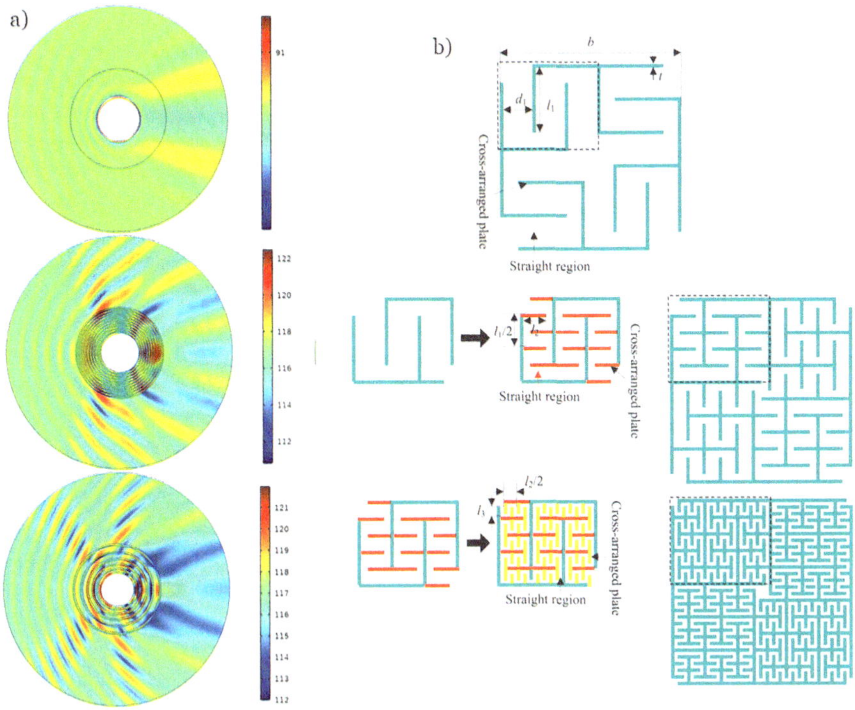

Fig. 4.1 (**a**) Acoustic cloak with visualized wave formation adapted from [22]. (**b**) Labyrinth acoustic cloak beginning with individual unit cell and full lattice formation from periodic translation adapted from [9]

vibrations [13]. Figure 4.3 depicts experimental results of an aluminum thermal cloak in the left column of images and a bare object in Fig. 4.3(i, ii, v, and vi) where the first two set of experiments contains the results of the heat flux cloaking capabilities of the material, and the second set contains the results of direct current field passing through the material [11].

4.2 Mechanisms of Cloaking and Invisibility

EIT principally works through quantum interference observed in classical three-level atomic systems, which results in an otherwise opaque medium becoming transparent in an extremely narrow window due to electromagnetic coupling between atoms. This coupling is the result of magnetic dipole/electric quadrupole resonance commonly called the dark mode couples with the electric dipole also known as the bright mode. Applications of EIT MMs include optics, wireless communications, low light, or slow-light devices [25–27]. To induce the EIT effect in these MMs, they are composed of periodic arrays of resonators of various

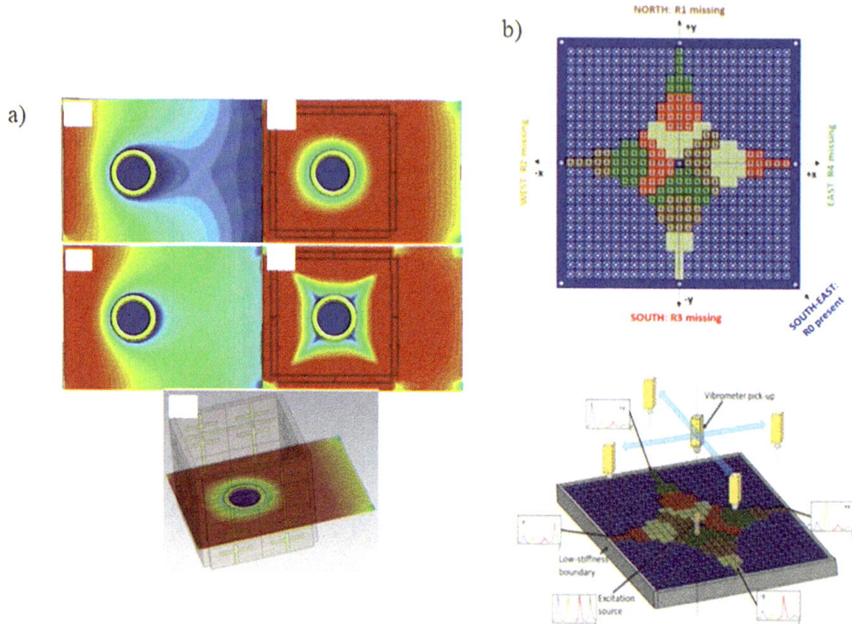

Fig. 4.2 (**a**) EM cloak from PCB arrangement of split resonator unit cells [23]. (**b**) Mechanical cloak with visualized deformation results adapted from [12]

geometries [28, 29]. These resonators are often made from polytetrafluoroethylene substrates with copper traces but do often rely on silicon, gold, graphene, and other high conductive materials to gather the optimal charge to produce the effect [30–32].

Figure 4.4 depicts several examples of EIT in action for differing MMs: (a) depicts the wave form propagation across the surface of an EIT electromagnetic MM in inactive and active states [33], (b) depicts the coupling modes required to produce the EIT effect [34], and (c) provides simulated results of the transmission spectrum across the surface of an EIT electromagnetic MM at different levels of Fermi energy [20]. A small but emerging area of focus in these materials is developing adaptive and tunable materials such that the level of transparency can be controlled and altered to fit the desired application, which is particularly suited for the infrared terahertz (THz) MMs as optical applications rely on the IR spectrum for visualization [35, 36].

Biomedical sensing is a popular application for EIT MMs. A novel medical study examined the possibility of using an EIT electromagnetic MM for the detection of cancer cells [37]. Their design was a simple hourglass shape double split ring resonator made from photolithography and composed of a silicon substrate, coated with a 25 μm-thick dielectric polyimide layer. This material was simulated and fabricated to validate experimentally and found that between 1.20 and 1.67 THz the electric field in-phase splits on both ends representing that region as the band for which EIT resonance occurs.

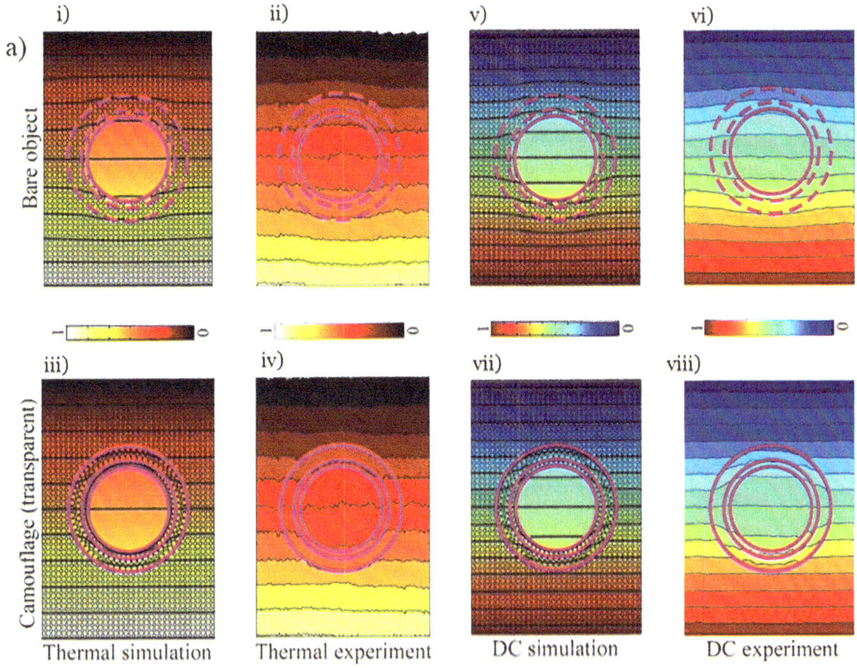

Fig. 4.3 Thermal cloak with experimentation (i–iv) and simulation (v–viii) visualization of the wave form where graphs (i), (ii), (v), and (vi) depict the bare object and graphs (iii), (iv), (vii), and (viii) depict the camouflaged object adapted from [10]

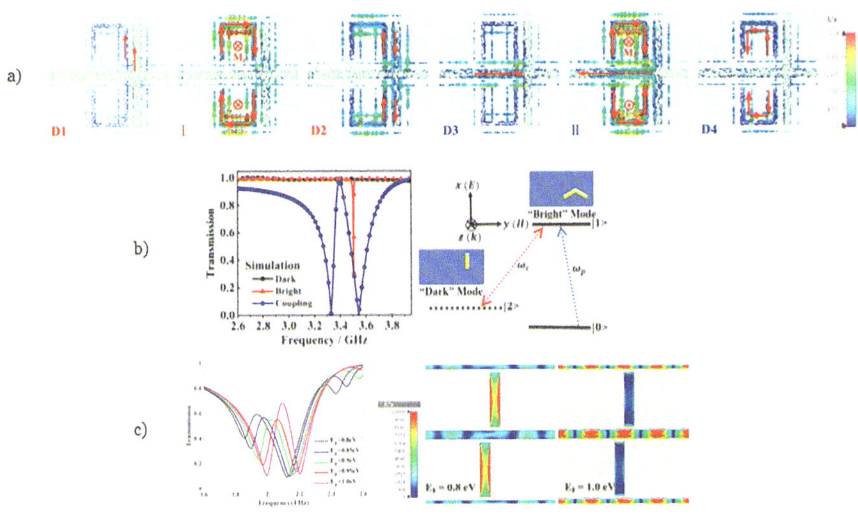

Fig. 4.4 Examples of EIT MMs. (**a**) Surface current wave propagation adapted from [30]. (**b**) Coupling mode of material to induce transparency adapted from [31]. (**c**) FEA simulation of EIT MM adapted from [19]

The Lorentz oscillator model, a classical phenomenological model that describes optical response of materials, is often used in this analysis as it provides an analytical model of an electron orbiting a massive stationary nucleus as a simple spring-mass-damper system [38, 39]. This study found the susceptibility χ through the use of the Lorentz model as a guide using the interference between two harmonic oscillators as the coupling effect between dark and light modes for a specific resonant frequency [36]:

$$\chi = \chi_r + i\chi_i \propto \frac{(\omega - \omega_0 + \delta) + i\frac{\gamma_2}{2}}{\left(\omega - \omega_0 + i\frac{\gamma_1}{2}\right)\left(\omega - \omega_0 - \delta + i\frac{\gamma_2}{2}\right) - \frac{\kappa_c^2}{4}} \tag{4.1}$$

The imaginary susceptibility χ_i was used to evaluate the transmissibility **J** as **J** = 1-**g** χ_i where ***g*** is a geometric parameter resultant from the double split ring resonator. This material was shown to have a maximum sensitivity of 900 kHz/cell ml^{-1} allowing for the detection of particularly small cancer cells fulfilling the desired experimental goals. Subsequent research built upon this work and demonstrated another EIT electromagnetic MM to detect malignant glioma cells [40]. Their electromagnetic MM fabricated through photolithography featuring four split ring resonators and cut wires made from chromium (10 nm thick layer) and gold (100 nm thick layer) deposited on a 500 μm thick quartz substrate. These metamaterials had a resulting higher permittivity than the earlier reported work leading to a higher peek frequency for EIT resonance of 2.24 THz. This material was shown to experience a red shift in simulation with increased refractive index. The maximum experimental sensitivity of this material was found to be 248.75 kHz/cell ml^{-1} which is well within a desirable range for detection of glioma cells.

4.3 Electromagnetic Cloaking

Permeability and permittivity are the properties utilized in defining the characteristics of an electromagnetic MM cloak. Here the wave frequency and transmissibility which the material is subjected to are used as a measure of performance to determine the best application of the material. Several of the material properties in electromagnetic MMs require specific methods to calculate [24].

Permittivity and permeability are properties that describe a material's ability to store electrical energy in response to an electric field, and the ability to support the formation of a magnetic field respectively. The scattering parameter method is a way to characterize the behavior of electrical networks at high frequencies by describing how incident waves are reflected and transmitted. The permittivity and permeability of the material can be calculated through the scattering parameter method by [23]:

$$\varphi = \frac{1}{k_0 d}\left(Im\left(\ln\left(\frac{1-aS_{11}}{S_{21}} \right) \right) + 2m\pi - i\,\mathrm{Re}\left(\ln\left(\frac{1-aS_{11}}{S_{21}} \right) \right) \right) \tag{4.2}$$

$$z = \pm\sqrt{\frac{(1+S_{11})^2 - S_{21}^2}{(1-S_{11})^2 - S_{21}^2}} \tag{4.3}$$

$$\in = \frac{n}{z} \tag{4.4}$$

$$\mu = nz \tag{4.5}$$

$$r = \frac{z-1}{z+1} \tag{4.6}$$

This approach yields the refractive index $\boldsymbol{\varphi}$, impedance z, permeability $\boldsymbol{\mu}$, absorption coefficient $\boldsymbol{a}$, and permittivity $\in$ of the material which are also important for determining the transmissibility and operating frequencies of the material. With the properties of the material determined, a composition can be chosen which fits the desired properties. Typically, in the fabrication on electromagnetic MM cloak, polytetrafluoroethylene circuit boards with copper or silicon contacts are typically chosen; however, on some occasions for higher frequency applications, gold or other conductive materials are used [41, 42]. The material selection and design of the SRR cause a large effect on the performance of the cloak. A unique electromagnetic MM cloaking study demonstrated a vanadium-dioxide metal based electromagnetic MM cloak capable of interacting with a terahertz band, between 0.5 and 0.7 THz for the purpose of being used as an optical device [6]. Among the materials surveyed, electromagnetic MM cloaking behavior was shown to be outclassed by a similar material that combines the properties of cloaking materials with the properties of absorbing materials known as transparent materials [43].

4.4 Acoustic Cloaking

In general, there are two approaches to take for acoustic cloaking MMs. The first is to create a lens to channel the acoustic waves through a medium, and the second is to use the geometry of the material to channel the acoustic waves around the structure of the object through mechanical means [23, 44].

Lens-type acoustic MM cloaks are more orderly and isotropic in their construction using a unit cell composed of resonators to vibrate in response to incident waves [45, 46]. These lenses are typically designed for detection of waves of specific frequencies [47, 48].

A spring-mass model is often useful in building the structure of the unit cells and their interactions with one another. The dynamic motion of the unit cells in response to the incident waves can be modeled for their natural frequency through dispersion relations. For these materials, the dispersion relation for the frequency of the lens can be obtained as [3]:

$$\omega = \sqrt{\frac{2\left(k_1 + k_n\right)\left(1 - \cos\left(k_{eff} a\right)\right)}{m_{eff}}} \tag{4.7}$$

Defining these parameters allows for the design of a helical cloaking material. Another form of lens uses a Luneburg lens to take a singular incident wave and split it into several beams scattering it around the material [49–51].

Mechanical design based acoustic MM cloaks, while typically labyrinthine in construction, utilize the geometry of the material to guide the waves across the medium. The working principle is that the sharp corners and long corridors of the material guide the waves along and around the material such that the vibrations resonate with and follow these paths [52, 53]. Since these materials are working off of the resonation of acoustic waves, the resulting motion and deformation of the material is able to be siphoned much like the purpose-made energy absorbing materials to create the piezoelectric effect for self-powered applications across low frequencies [54, 55]. Figure 4.5 provides an example of the pressure concentrations and eigenstates for a third-order labyrinthine acoustic cloak with accompanying transmission coefficient and sound transmission loss (STL) diagrams demonstrating the cloaking effect impeding transmission of sound waves within the frequency range of 400–800 Hz [53].

4.5 Thermal Cloaking

Thermal cloaking is a popular form of thermal MM developed using finite element analysis to simulate the performance of the material [56]. These materials are primarily developed for medical imaging, particularly in cancer detection by recognizing the unique heat signature of tumors and other large growths in a patient's body [57]. Most thermal MMs are designed for room temperature operating conditions, where conduction is the primary mode of heat transfer. This comes with its own advantages allowing for programmable properties through subtly tuning the thermal conduction coefficient of the material [58].

Fourier's law governing heat conduction is a fundamental law describing the rate of heat transfer being proportional to the negative temperature gradient and area perpendicular to the gradient through which heat flows. For metamaterials this can be used to design a lattice structure of a desired area to direct the flow of heat across a prescribed gradient. Hybrid tunable thermal MM cloaks work on the Fourier's laws of heat conduction without a heat source as described by the following [59]:

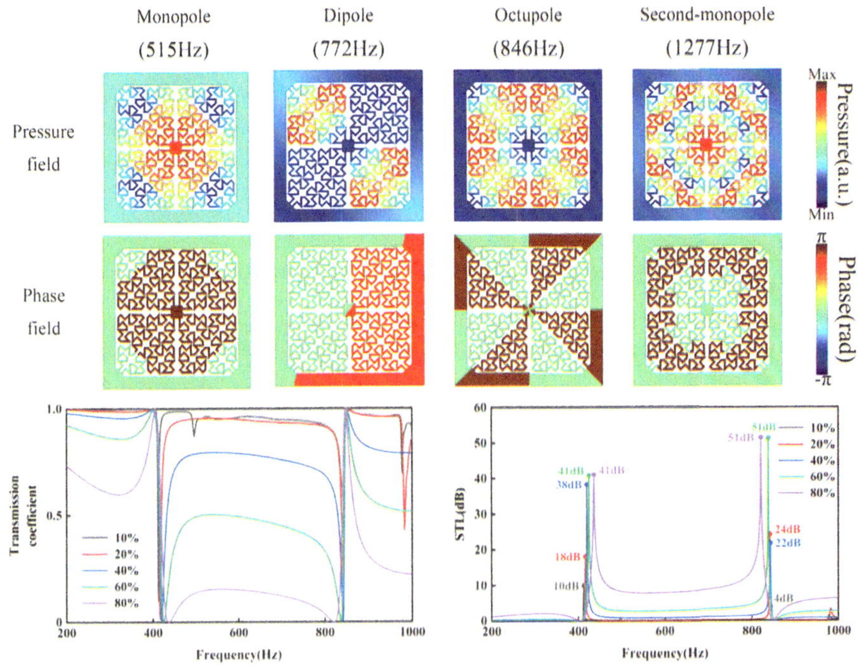

Fig. 4.5 Eigen states of a single unit third-order labyrinth acoustic cloak, corresponding transmission curves for fill rates from 10% to 80%, and STL for fill rates from 10% to 80% for the same cloak adapted from [49]

$$\rho C \frac{\partial T}{\partial t} = \nabla \cdot (\kappa \cdot \nabla T) \tag{4.8}$$

These materials are constructed from two layers that when aligned properly produce the cloaking effect. The rotation of these materials in and out of phase creates the tunable nature of the thermal MM cloak [11, 60, 61]. Figure 4.6 highlights an example of thermal cloaking metamaterials. Figure 4.6a illustrates the experimental set up and results of a thermal cloak thermal MM, which uses shape memory through the application of heat to encode desired functionality onto the material. This thermal MM is designed to control the flow of heat around the central region either to restrict heat from passing through or to concentrate heat through it; this switchable behavior can be observed in diagrams Fig. 4.6(iii, iv) where the heat is either directed through region or restricted from it as the blue line [59].

Figure 4.7 illustrates another cloaking thermal MM comprising a material which manipulates through the application of Flick's second law of diffusion, which describes the rate of accumulation of a concentration within a volume proportional to the curvature of the concentration gradient. This material achieves its desired effect through concentric external layers of different base materials consisting of various rubbers. This creates multiple functional regions of differing thermal

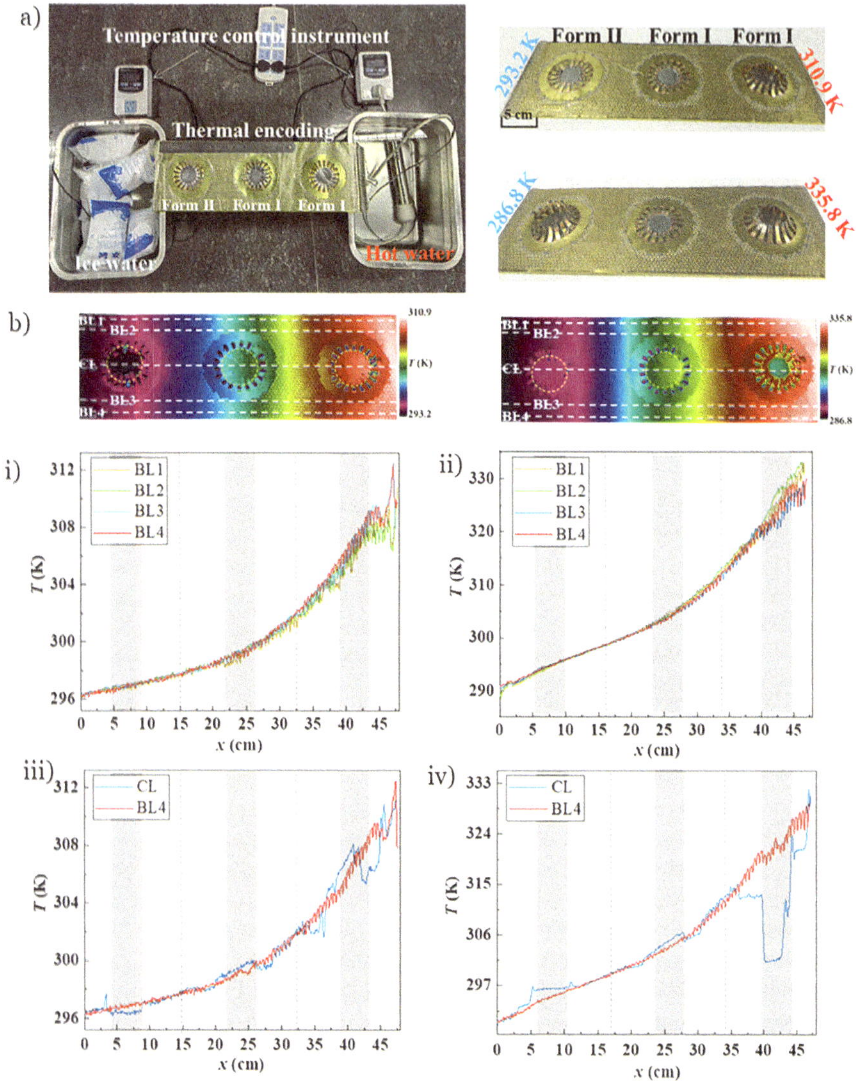

Fig. 4.6 Experimental set up (**a**), with results at different temperatures with temperature distribution diagram under different calibrations (**b**). Charts illustrate temperature distribution in the background region (i, ii) where charts (iii), and (iv) provide comparison between background and central regions adapted from [55]

conductivity that will allow the switching behavior to occur. This behavior cloaks the inner region from dioxide (O_2; Fig. 4.7a) in one state and converging dinitride (N_2; Fig. 4.7b) through the material in the other [60]. Figure 4.7c, e denotes the concentration of O_2 along the *x* and *y* axes, respectively. Similarly, Fig. 4.7d, f provides the concentration of N_2 along the *x* and *y* axes. From these two sets of graphs,

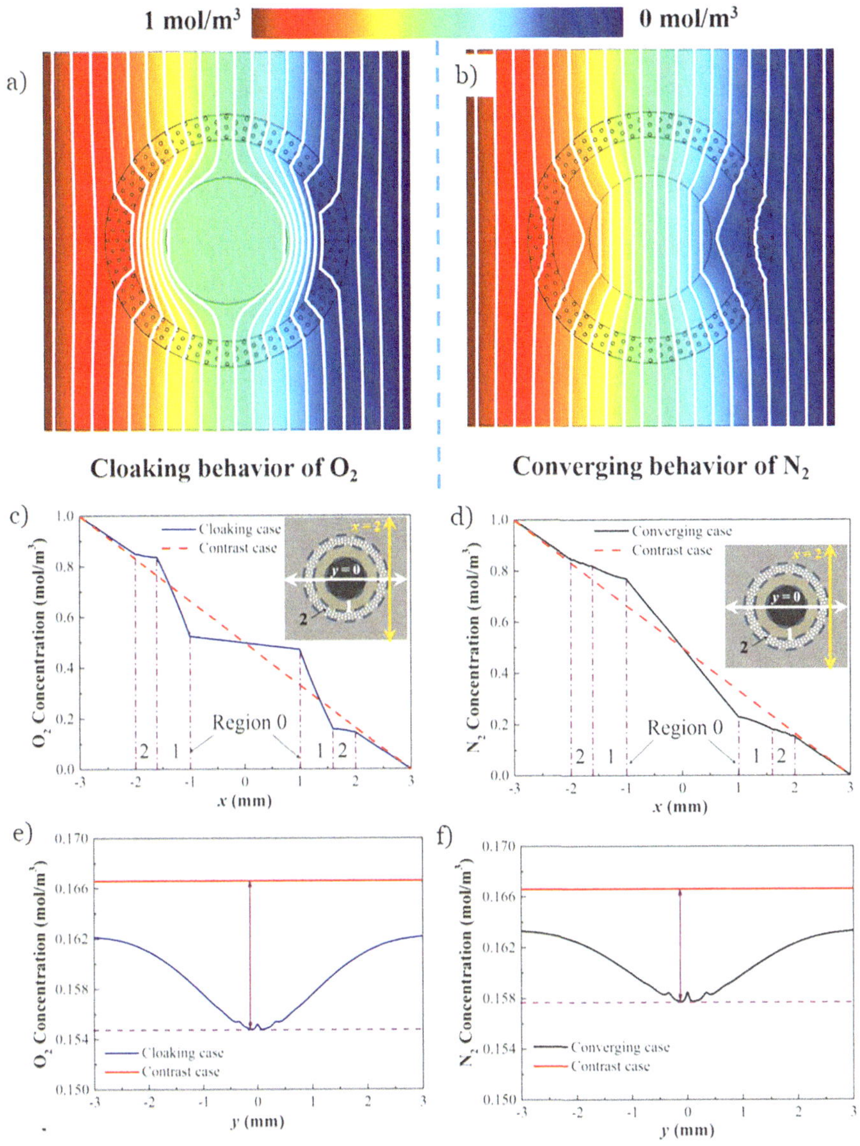

Fig. 4.7 Concentration fields for thermal cloaking metamaterial demonstrating cloaking behavior under O_2 (**a**), and convergent behavior under N_2 (**b**). Diagrams depict O_2 concentration fields along $y = 0$ mm (**e**), and $x = 2$ mm respectively (**c**), and corresponding N_2 concentrations along $y = 0$ mm (**f**) and $x = 2$ mm (**d**) respectively adapted from [56]

the switching behavior can be clearly observed as in either direction over the length of the material O_2 is unable to permeate the center and N_2 is able to pass virtually unimpeded.

4.6 Summary

The sophisticated design of metamaterials tailored for manipulating wave phenomena, with a specific focus on cloaking has been discussed. By tailoring the geometry of the unit cells, these materials can control local resonance effectively, allowing them to guide waves around objects, thereby rendering them functionally invisible. This ability to manipulate wave propagation opens a host of promising applications across various fields. In defense, cloaking MMs offer the potential to develop equipment and vehicles that are invisible to radar and sonar, providing a strategic advantage that could redefine modern warfare. Similarly, in the civilian realm, these materials have significant implications for noise cancellation technologies, potentially leading to quieter living environments by mitigating noise pollution without the need for physical barriers.

The medical field will also benefit from cloaking MMs. In medical imaging, acoustic cloaks can enhance the clarity and precision of ultrasound and other imaging techniques, allowing for better diagnosis and treatment planning without the use of invasive exploratory surgeries. Additionally, the potential applications in earthquake detection and structural protection suggest a future where buildings and infrastructure can be designed to automatically mitigate damage from seismic activity, vastly improving safety and resilience.

The ongoing advancements in the research and development of cloaking metamaterials are expected to push the boundaries of what is currently achievable, expanding the capabilities of these materials to encompass a broader range of frequencies and applications. As researchers continue to unlock new ways to manipulate wave interactions through advanced material design, the next generation of metamaterials could transform our ability to control and utilize several types of wave energy, paving the way for innovative applications that could profoundly impact society and industry.

References

1. Xu, Q., Liu, X., Xuan, Y., Xu, Y., & Liu, D. (2020). High-performance infrared thermal radiation suppression metamaterials enabling inhibited infrared emittance and decreased temperature simultaneously. *International Journal of Heat and Mass Transfer, 161*, 120318.
2. Chen, Z., Zhou, W., & Lim, C. W. (2021). Tunable frequency response of topologically protected interface modes for membrane-type metamaterials via voltage control. *Journal of Sound and Vibration, 494*, 115870.
3. Li, X., Li, J., & Huang, X. (2023). Reconfigurable coding acoustic meta-lens based on helical metamaterials. *Applied Acoustics, 211*, 109538.
4. Fang, Z., Li, J., & Wang, X. (2020). Optimal control for electromagnetic cloaking metamaterial parameters design. *Computers & Mathematics with Applications, 79*(4), 1165–1176.
5. Gintides, D., Pallikarakis, N., & Stratouras, K. (2021). On the modified transmission eigenvalue problem with an artificial metamaterial background. *Research in the Mathematical Sciences, 8*(3), 40.

6. Wang, D., Tian, J., Ma, T., Zhang, Y., & Yue, S. (2024). Terahertz modulators based on VO2–metal hybridized metamaterials for free switching between BIC and quasi-BIC states. *Optics Communications, 551*, 130040.
7. Nanda, S., Sahu, P. K., & Mishra, R. K. (2019). FLANN-SM analysis of a $\varOmega$ unit cell. *Journal of Computational Electronics, 18*(3), 1007–1016.
8. Nguyen, N. Q., & Gupta, N. (2008). Optical fiber loop sensors for structural health monitoring of composites. *MRS Proceedings, 1129*, 1129-V07-11.
9. Zhou, W., Su, Y., Muhammad, Chen, W., & Lim, C. W. (2020). Voltage-controlled quantum valley Hall effect in dielectric membrane-type acoustic metamaterials. *International Journal of Mechanical Sciences, 172*, 105368.
10. Liu, J., Li, L., Xia, B., & Man, X. (2018). Fractal labyrinthine acoustic metamaterial in planar lattices. *International Journal of Solids and Structures, 132-133*, 20–30.
11. Zhang, X., He, X., & Wu, L. (2021). A bilayer thermal-electric camouflage device suitable for a wide range of natural materials. *Composite Structures, 261*, 113319.
12. Long, J., Wang, A., Zhou, Y., & Chen, C. Q. (2023). Quadrupole higher-order topological phases in static mechanical metamaterials. *International Journal of Mechanical Sciences*, 108782.
13. Emerson, T. A., & Manimala, J. M. (2020). Passive-adaptive mechanical wave manipulation using nonlinear metamaterial plates. *Acta Mechanica, 231*(11), 4665–4681.
14. Galdetskii, A. V. (2022). On the possibility of using a metamaterial to increase the power of multibeam klystrons. *Journal of Communications Technology and Electronics, 67*(10), 1224–1230.
15. Feng, S. N. A., Wang, Y., Fei, S., Yan, Z., Yu, L., Chen, J., Tang, C., & Liu, F. (2023). Dual ultrahigh-Q Fano resonances of 3D gap metamaterials for slow light from ultraviolet to visible range. *Optics Communications, 549*, 129811.
16. Yang, S., Xu, L., Dai, G., & Huang, J. (2020). Omnithermal metamaterials switchable between transparency and cloaking. *Journal of Applied Physics, 128*(9).
17. Askari, M., & Bahadoran, M. (2022). A refractive-index-based microwave sensor based on classical electromagnetically induced transparency in metamaterials. *Optik, 253*, 168589.
18. Zhang, S., Qin, M., Wu, B., & Wu, E. (2023). All-dielectric Si metamaterials with electromagnetically induced transparency and strong gap-mode electric field enhancement. *Optics Communications, 530*, 129143.
19. Zhang, C., Wang, Y., Yao, Y., Tian, L., Geng, Z., Yang, Y., Jiang, J., & He, X. (2020). Active control of electromagnetically induced transparency based on terahertz hybrid metal-graphene metamaterials for slow light applications. *Optik, 200*, 163398.
20. Liu, C., Liu, P., Bian, L., Zhou, Q., Li, G., & Liu, H. (2018). Dynamically tunable electromagnetically induced transparency analogy in terahertz metamaterial. *Optics Communications, 410*, 17–24.
21. Mo, Y., Zhong, J., & Lin, Y.-S. (2020). Tunable chevron-shaped infrared metamaterial. *Materials Letters, 263*, 127291.
22. Keshavarz, A., & Zakery, A. (2018). A novel terahertz semiconductor metamaterial for slow light device and dual-band modulator applications. *Plasmonics, 13*(2), 459–466.
23. Zaremanesh, M., & Bahrami, A. (2022). Multilayer acoustic invisibility cloak based on composite lattice. *Scientific Reports, 12*(1), 16096.
24. Zhao, G., Bi, S., Niu, M., & Cui, Y. (2019). A zero refraction metamaterial and its application in electromagnetic stealth cloak. *Materials Today Communications, 21*, 100603.
25. Cai, W., Xiao, B., Yu, J., & Xiao, L. (2020). A compact graphene metamaterial based on electromagnetically induced transparency effect. *Optics Communications, 475*, 126266.
26. Shen, Z.-Y., Yang, H.-L., Liu, X., Huang, X.-J., Xiang, T.-Y., Wu, J., & Chen, W. (2019). Electromagnetically induced transparency in novel dual-band metamaterial excited by toroidal dipolar response. *Frontiers of Physics, 15*(1), 12601.
27. Zhang, Y., Feng, Y., & Zhao, J. (2020). Graphene-enabled tunable multifunctional metamaterial for dynamical polarization manipulation of broadband terahertz wave. *Carbon, 163*, 244–252.

28. Chen, M., Xiao, Z., Lu, X., Lv, F., & Zhou, Y. (2020). Simulation of dynamically tunable and switchable electromagnetically induced transparency analogue based on metal-graphene hybrid metamaterial. *Carbon, 159*, 273–282.
29. Ma, Y., Huang, L., & Chen, L. (2023). Active control of broadband plasmon induced transparency window and amplitude modulation based on embedded-vanadium-dioxide metamaterials. *Optics Communications, 545*, 129706.
30. Li, R., Kong, X. K., Liu, S. B., Liu, Z. M., & Li, Y. M. (2019). Planar metamaterial analogue of electromagnetically induced transparency for a miniature refractive index sensor. *Physics Letters A, 383*(32), 125947.
31. Deng, Y., & Song, Z. (2020). Manipulating polarization and electromagnetically induced transparency in a switchable metamaterial. *Optical Materials, 105*, 109972.
32. Liu, C., Liu, P., Yang, C., Lin, Y., & Liu, H. (2019). Analogue of dual-controlled electromagnetically induced transparency based on a graphene metamaterial. *Carbon, 142*, 354–362.
33. Chen, M., & Xiao, Z. (2022). Metal-graphene hybrid terahertz metamaterial based on dynamically switchable electromagnetically induced transparency effect and its sensing performance. *Diamond and Related Materials, 124*, 108935.
34. Du, C., Zhou, D., Guo, H.-H., Pang, Y.-Q., Shi, H.-Y., Liu, W.-F., Singh, C., Trukhanov, S., Trukhanov, A., & Xu, Z. (2021). Active control scattering manipulation for realization of switchable EIT-like response metamaterial. *Optics Communications, 483*, 126664.
35. Liu, T., Wang, H., Liu, Y., Xiao, L., Yi, Z., Zhou, C., & Xiao, S. (2018). Active manipulation of electromagnetically induced transparency in a terahertz hybrid metamaterial. *Optics Communications, 426*, 629–634.
36. Chen, M., Xiao, Z., Lv, F., Cui, Z., & Xu, Q. (2020). Dynamically tunable dual-band electromagnetically induced transparency-like in terahertz metamaterial. *Optical Materials, 107*, 110060.
37. Yan, X., Yang, M., Zhang, Z., Liang, L., Wei, D., Wang, M., Zhang, M., Wang, T., Liu, L., Xie, J., & Yao, J. (2019). The terahertz electromagnetically induced transparency-like metamaterials for sensitive biosensors in the detection of cancer cells. *Biosensors and Bioelectronics, 126*, 485–492.
38. AbouelNour, Y., & Gupta, N. (2023). Assisted defect detection by in-process monitoring of additive manufacturing using optical imaging and infrared thermography. *Additive Manufacturing, 67*, 103483.
39. Nishino, Z. T., Chen, K., & Gupta, N. (2014). Power modulation-based optical sensor for high-sensitivity vibration measurements. *IEEE Sensors Journal, 14*(7), 2153–2158.
40. Zhang, J., Mu, N., Liu, L., Xie, J., Feng, H., Yao, J., Chen, T., & Zhu, W. (2021). Highly sensitive detection of malignant glioma cells using metamaterial-inspired THz biosensor based on electromagnetically induced transparency. *Biosensors and Bioelectronics, 185*, 113241.
41. Murai, N., Noguchi, Y., Matsushima, K., & Yamada, T. (2023). Multiscale topology optimization of electromagnetic metamaterials using a high-contrast homogenization method. *Computer Methods in Applied Mechanics and Engineering, 403*, 115728.
42. Nguyen, N. Q., & Gupta, N. (2009). Analysis of an encapsulated whispering gallery mode micro-optical sensor. *Applied Physics B, 96*(4), 793–801.
43. Peng, Y., Wang, Q., Xu, Y., Shan, D., He, L., & Cao, Y. (2023). Optically transparent and mechanically stretchable fractal-structured wave-absorbing metamaterial in low frequency range. *Journal of Alloys and Compounds, 961*, 171100.
44. Madeo, A., Collet, M., Miniaci, M., Billon, K., Ouisse, M., & Neff, P. (2018). Modeling phononic crystals via the weighted relaxed micromorphic model with free and gradient micro-inertia. *Journal of Elasticity, 130*(1), 59–83.
45. Jeon, J. Y., Hong, S. H., Choi, E. Y., Bok, E., Choi, S. H., & Seo, J. W. (2021). Intuitive understandings of negative bulk modulus of metamaterials composed of Helmholtz resonators. *Current Applied Physics, 29*, 128–132.
46. Chen, P., Haberman, M. R., & Ghattas, O. (2021). Optimal design of acoustic metamaterial cloaks under uncertainty. *Journal of Computational Physics, 431*, 110114.

47. Huang, K.-X., Shui, G.-S., Wang, Y.-Z., & Wang, Y.-S. (2020). Meta-arrest of a fast propagating crack in elastic wave metamaterials with local resonators. *Mechanics of Materials, 148*, 103497.
48. Piatysheva, A. P., & Grekova, E. F. (2021). Reduced linear viscoelastic isotropic Cosserat medium with rotational viscosity: An acoustic metamaterial. *Continuum Mechanics and Thermodynamics, 33*(4), 1765–1780.
49. Zhao, L., Sun, Z., Liu, Q., & Bi, C. (2023). Acoustic beam splitter based on acoustic metamaterial Luneburg lens. *Physics Letters A, 472*, 128815.
50. Eslamzadeh, S., Ghaffari-Miab, M., & Abbasi-Arand, B. (2021). Design of a broadband metamaterial-based acoustic lens using elaborated carpet cloak strategy. *Applied Physics A, 127*(12), 897.
51. Lee, J. J. (2020). A mixed method for time-transient acoustic wave propagation in metamaterials. *Journal of Scientific Computing, 84*(1), 20.
52. Huang, R.-n., Xin, Y.-j., Cheng, S.-l., Sun, Y., Yan, Q., & Ding, Q. (2023). Scattering properties of labyrinthine metamaterials with different numbers of resonance units. *Physica B: Condensed Matter, 671*, 415446.
53. Xin, Y.-j., Huang, R.-n., Li, P., Qian, Q., Yan, Q., Sun, Y.-t., Ding, Q., Huang, Y.-j., & Cheng, S.-l. (2023). Labyrinthine acoustic metamaterials with triangular self-similarity for low-frequency sound insulation at deep subwavelength dimensions. *Results in Physics, 54*, 107151.
54. Gao, W., Hu, J., Qin, Z., & Chu, F. (2023). Flexural wave manipulation in perforated metamaterial plates with acoustic black holes interconnected by piezoelectric studs. *Composite Structures, 321*, 117224.
55. Xue, Y., & Zhang, X. (2021). Self-adaptive acoustic cloak enabled by soft mechanical metamaterials. *Extreme Mechanics Letters, 46*, 101347.
56. Yu, Z., Ren, Z., & Lee, J. (2019). Investigation of thermal metamaterials based on nanoporous silicon using ray tracing and finite element simulations. In *2019 18th IEEE Intersociety Conference on Thermal and Thermomechanical Phenomena in Electronic Systems (ITherm)*.
57. Islam, M. S., Islam, M. T., & Almutairi, A. F. (2022). A portable non-invasive microwave based head imaging system using compact metamaterial loaded 3D unidirectional antenna for stroke detection. *Scientific Reports, 12*(1), 8895.
58. Wang, Z., Liu, Z., & Lee, J. (2020). Tuning the working frequency of elastic metamaterials by heat. *Acta Mechanica, 231*(4), 1477–1484.
59. Lei, M., Jiang, C., Yang, F., Wang, J., & Huang, J. (2023). Programmable all-thermal encoding with metamaterials. *International Journal of Heat and Mass Transfer, 207*, 124033.
60. Zhou, X., Xu, G., & Zhang, H. (2021). Binary masses manipulation with composite bilayer metamaterial. *Composite Structures, 267*, 113866.
61. Jin, P., Liu, J., Xu, L., Wang, J., Ouyang, X., Jiang, J.-H., & Huang, J. (2023). Tunable liquid–solid hybrid thermal metamaterials with a topology transition. *Proceedings of the National Academy of Sciences, 120*(3), e2217068120.

Chapter 5
Design and Manufacturing

List of Symbols

α_{ij}^{e}	Local thermal expansion coefficient vector
$\boldsymbol{E}_{ijkl}^{e}$	Local elasticity matrix of element E
$\boldsymbol{P}_{\text{cons}}$	Expected cell value
$\boldsymbol{X}_{e}'$	Adjacent cell values

5.1 Introduction

The manufacturability of a metamaterial (MM) is partially determined by its functionality. After running simulations on a material to determine the desired properties, a microstructure and material composition can be chosen. For electromagnetic MMs electrochemical processes are appropriate for fabrication as silicon, graphene, and copper are common elements found in electromagnetic MMs and traditional printed circuit board manufacturing can be employed from there to form the material. While other forms of manufacturing such as subtractive or formative have a place in the fabrication of MMs, additive manufacturing allows creating complex geometries and features that these traditional methods are unable to produce. Vat polymerization, material extrusion, and powder bed fusion based additive manufacturing methods have been used to produce metamaterials. While all forms of additive manufacturing are used to create metamaterials, those are the most popular due to their availability, material selection, and resolution [1–3]. Additive manufacturing can produce materials with complex geometries; each process has strict limitations to the level of resolution that can be produced. Metamaterials require the production of lattices at a resolution below what is capable of being the most commercially available 3D printers to properly observe their interaction media. Hence, more specialized processes are used to construct these materials such as single crystals, utilizing patterns to achieve the correct shape, as well as two photon polymerization methods of additive manufacturing. Often entirely new material does not

N. Gupta, C. Beckwith, *Architected Metamaterials: Design Principles and Properties*, https://doi.org/10.1007/978-3-031-98330-6_5

need to be constructed, and rather an assembly approximating the geometric conditions of the material can be created and used to verify theoretical simulations. The manufacturing requires specialized techniques to quantify their compositions and ensure the accuracy of their manufacturing process [4–6]. These assemblies have been made from machined metal parts, cut and folded cardboard slats, and even popsicle sticks pinned together in the desired shape [7].

5.2 Topology Optimization

Topology optimization represents an evolutionary leap in material design, leveraging algorithms to iteratively refine an object's shape based on predetermined material properties. Among the various methods employed, the solid isotropic material with penalization (SIMP) technique is particularly notable. SIMP methodically reduces an object's mass to meet specific load cases, iteratively eliminating material, while ensuring a specific minimum safety factor is met [8–10]. Although effective for broad applications, SIMP may not fully capture complex property requirements of advanced MMs. Consequently, research has expanded toward developing lattice structures from simple unit cells, capable of embodying a more diverse array of material properties, thereby facilitating the creation of innovative MMs [11–13].

A practical application of topology optimization is observed in the aerospace industry, where minimizing weight while preserving structural integrity is paramount. For example, the design of an aircraft engine bracket can be optimized to reduce the material in areas experiencing negligible levels of stress based on the simulations. This not only results in a lighter component but also maintains or enhances the product's strength and durability. Unlike conventional design methods such as parametric modeling that may rely on heuristic approaches, topology optimization employs a systematic, algorithm-driven process that can explore a vast design space more efficiently. This allows for the discovery of unconventional design solutions such as intricate lattice structures for stress distribution, which might not be intuitive through traditional design methodologies.

Machine learning (ML) has further enhanced topology optimization capabilities, enabling the analysis of vast datasets to predict material behaviors under varied conditions [14–16]. For example, algorithms can now design cloaking metamaterials that deviate from the traditional cylindrical shape, tailored to specific electromagnetic wave interactions [17–19]. A notable advancement is the development of a genetic and evolutionary ML models that constructs fractal lattices based on desired stress-strain curves, enriching the design process with data-driven, physics-informed insights [20, 21]. Challenges such as data redundancy and the risk of overfitting necessitate continuous refinement of these models. Figure 5.1 depicts the topology optimization process chain of an evolutionary strategy optimization algorithm developed from a large library of stress-strain curves to complete assessment of optimized lattice structures [17]. The process chain begins at (a) with training data, consisting of thousands of stress-strain curves being compiled and fed into the

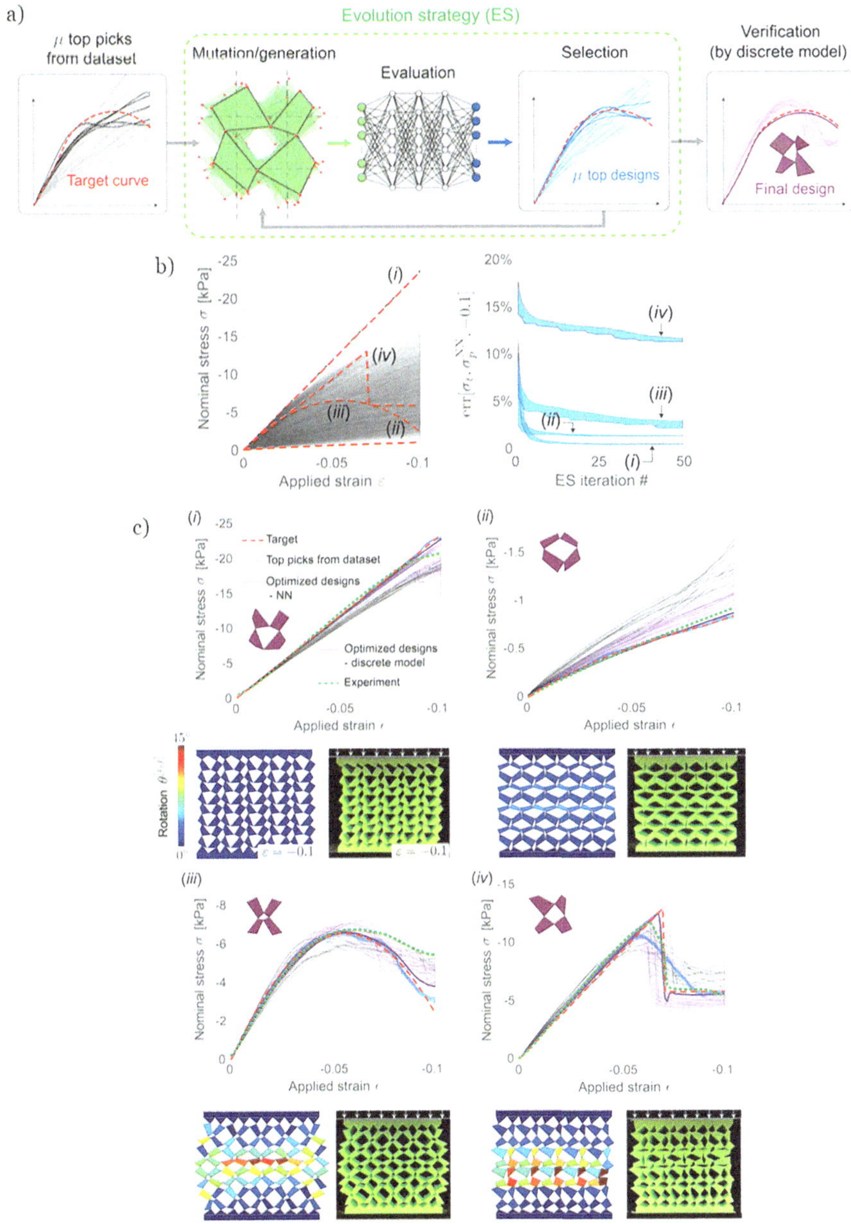

Fig. 5.1 MM topology optimization of fractal lattice: (**a**) Training dataset stress strain cures fed into evolutionary strategy deep learning algorithm with optimized model parameter outcome. (**b**) Resultant targeted stress curves marked in red with evolutionary iteration chain of the considered target material properties for the desired geometries. (**c**) Targeted response from the optimized designed from resultant dataset in the upper pictures (i–iv), lower pictures include the finite element simulations and printed compression samples for target curves (i–iv) adapted from [17]

evolutionary strategy-based optimization model. The next step (b) picks out target curves which are then iterated such that their geometry matches the desired curves. The last step (c) is to simulate and validate the performance of the optimization algorithm by fabricating the lattices and evaluating them through compression testing. These curves are then fed back into the training set to further refine the model.

An innovation within topology optimization is the polynomial-chaos-Chebyshev-interval (PCCI) method, aimed at engineering thermal MMs with exceptional thermoelastic properties [22]. By delving into the microstructure, the PCCI method evaluates how variations in material composition influence the overall mechanical and thermal behavior, highlighting the iterative nature of topology optimization in refining material geometry for optimal performance.

Modeling the MM mathematically often includes breaking the material down into its constituent components such as the case of the PCCI material where it uses two compositions locally to inform the macro performance of the material. By modeling the material across its boundary domain, the properties of the two materials can be summed to determine its overall performance. The requirements this method looks to resolve within the microstructure of the material can be expressed as [19]:

$$\boldsymbol{E}^{e}_{ijkl} = \left(x^{e}\right)^{p} \cdot \left(\left(y^{e}\right)^{p_2} \cdot \boldsymbol{E}^{2}_{ijkl} + \left(1 - y^{e}\right)^{p_2} \cdot \boldsymbol{E}^{1}_{ijkl}\right) \tag{5.1}$$

$$\alpha^{e}_{ij} = \left(y^{e}\right)^{p_2} \cdot \alpha^{2}_{ij} + \left(1 - y^{e}\right)^{p_2} \cdot \alpha^{1}_{ij} \tag{5.2}$$

where $\boldsymbol{E}^{e}_{ijkl}$ and α^{e}_{ij} are the local elasticity matrix and the local strain tensor for the element e. these properties are put through a sensitivity analysis which follows the flow of Fig. 5.2, where $\boldsymbol{P}_e$ the perimeter of the element within the design domain is

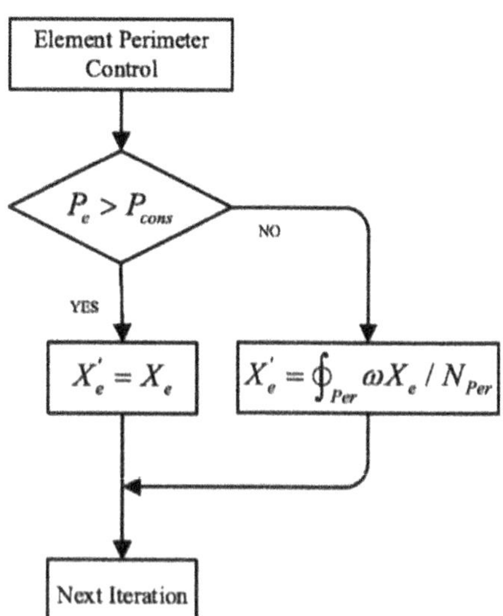

Fig. 5.2 Unit cell iteration for thermoelastic thermal MM adapted from [19]

calculated to the design variables and expected values $\mathbf{P}_{\mathbf{cons}}$. If $\boldsymbol{P}_{\mathbf{cons}}$ is greater than $\boldsymbol{P}_e$ then that unit cell is filtered against the adjacent cells as $\boldsymbol{X}_e'$, which is used for the next iteration in the cycle. This process is exemplified in Fig. 5.2, showing that the lattice topology will continue to iterate. Similar methods utilize other material properties such as Poisson's ratio and thermal expansion or base their machine learning algorithm on genetic models as they take simplified expressions for the material properties and structural behaviors to create reliable nonlinear systems [23, 24].

Carbon nanotubes and layered composites are commonly used in the development of metamaterials; however, the material properties of laminates are difficult to predict with respect to strain rates [25, 26]. ML can be used to develop predictive algorithms to determine the properties of these types of materials. A novel method was developed to predict these material properties, trained on viscoelasticity data gathered from dynamic mechanical analysis testing. By applying a superposition over the frequency and temperature of the laminate, a master curve can be generated at multiple strain rates to determine the Young's modulus of the material with respect to strain rates. Evaluating it at multiple angles ranging from 0° to 90° creates a map of not only the master-curves of the material properties but also the properties across anisotropic boundary conditions. The structure of this neural network and how it layered the extracted data into final curves can be seen in Fig. 5.3 as the temperature and frequency data are the input parameters and the time domain elastic properties are the ultimate output as depicted in the graph ranging between 0°, 45°, and 90° [27].

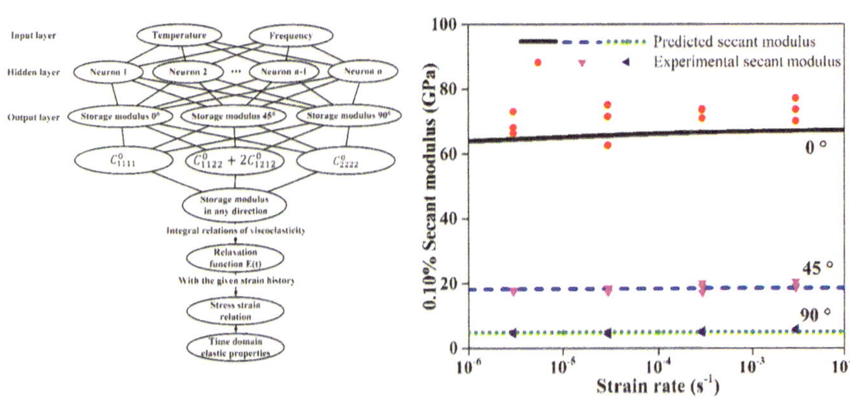

Fig. 5.3 Neural network structure and resultant prediction for secant modulus of carbon fiber laminate structure evaluated at 0°, 45°, and 90° adapted from [24]

5.3 Electrochemical Fabrication

Electrochemical fabrication emerges as a pivotal technique in the manufacturing of MMs, leveraging electrochemical reactions to precisely deposit or remove material on a substrate. This method's capacity to operate at the nanometer scale is particularly beneficial for MMs, facilitating the production of components like circuit boards and antennas that are integral for high-frequency wave interaction in applications such as optical microscopy and fiberoptic communications. An illustration of electrochemical fabrication's utility is in the production of high-precision antennas for terahertz (THz) communications. Through anodic deposition, a thin, conductive gold layer can be meticulously formed on a silicon substrate. This layer acts as an antenna, engineered to specific dimensions to resonate at THz frequencies, this enhancing signal reception and transmission in advanced communication devices. Electrochemical fabrication distinguishes itself from other methods like lithography or physical vapor deposition by its ability to directly "grow" materials onto a substrate, allowing for more complex three-dimensional structures, while lithography excels in patterning surface layers, electrochemical fabrication provides a depth of material deposition that is crucial for functional MMs requiring precise dimensional control and accuracy.

Figure 5.4 exemplifies several of the most common methods of electrochemical fabrication [28]. Similarly, each method utilizes differing electrochemical properties. (a) Anodic deposition is a controlled process of growing a thin oxide layer on a material such as aluminum or titanium, dissolving that in an electrolyte solution, releasing the metal ions, and forming metal-organic frameworks (MOFs) on the anode substrate. (b) Cathodic deposition similarly uses the cathode reaction to generate metal-organic frameworks on the substrate by metal ions in a similar electrolyte solution. (c) The third process converts precursor formations into MOFs by either hydrothermal or solvothermal methods. (d) Electrophoretic deposition deposits charged MOF particles on the electrode by driving an electric field. (e) Lastly, galvanic displacement reduces one metal on a substrate, while the substrate itself is oxidized, because of the differing potentials metal ions that are generated and form MOF films in the solution when brought into contact with organic linkers.

These processes become relevant to metamaterials that require very small sized structures for high-frequency applications such as electromagnetic absorption and transparent electromagnetic MMs, many terahertz applications also require some form of electroplating to properly dope their material [29–31]. MMs are composed of composite materials requiring high-temperature sintering through pyrolysis after electrochemical processing to reach their final state [32–34]. Vanadium dioxide is a powerful compound with electric properties exceeding that of silicon, making it extremely popular for metamaterials for THz applications. Hence, developing electrochemical processing parameters for MMs is also an area of focus in the design space [35–37].

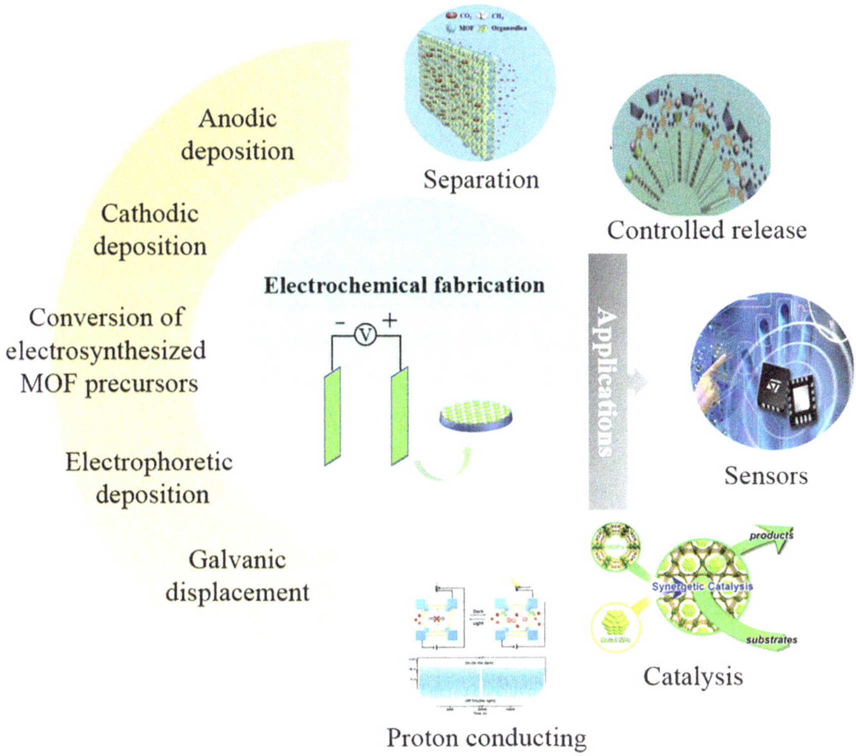

Fig. 5.4 Graphical representation of the most common electrochemical fabrication methods adapted from [25]

5.4 Additive Manufacturing

Additive manufacturing has emerged as a powerful method of manufacturing MMs. Additive manufacturing (AM) is a process of building up an object by depositing layers on top of one another. This allows for the creation of extraordinarily complex and asymmetric geometries which can perform the functionality desired by MMs. Material extrusion, vat polymerization, and powder bed fusion are the most popular variations of AM for production of MMs as they allow for the widest range of material types to fabricate with, as well as the highest resolutions for most commercially available printers [38, 39].

Material extrusion has become a popular method of producing MMs of all types due to the simplicity of this method for polymers such as polylactic acid (PLA), acrylonitrile butadiene styrene (ABS), and thermoplastic polyurethanes (TPU). Ceramics and metals have been printed through this method by creating a filament of powder of the modeling material mixed in a polymer. This process is extremely versatile, creating highly complex parts that machining or casting processes would be unable to manufacture, which is why it has become so popular for creating metamaterials [40, 41]. Material extrusion method was used to print an energy absorbing acoustic MM that acted as a siphon for low frequency broadband applications [42]. This was done to reduce the thickness of membrane type MMs to optimize the structure of the material without sacrificing performance. The unit cells were created through a material extrusion process, filled with silicon, and topped with aluminum platelet resonators such that the silicon would deform in response to the incident wave and the aluminum would absorb the sound and trap it between half of the resonator. Auxetic and energy absorbing metamaterials are commonly produced through material extrusion as the deformation in plastics is more easily observed from materials with low stiffness such as plastics rather than metals [43, 44]. Figure 5.5 provides insight into the simulation and performance of printed chiral-auxetic lattice structures at different applied displacements [44]. Herein the elongation pattern of the chiral metamaterial is exposed, revealing the reinforcement (Fig. 5.5b) added to the chiral pattern increases the average stress on the lattice as it is stretched. Due to the reinforcements Poisson's ratio of the structure is visibly decreased from Fig. 5.5a, b. Both lattices have an initially negative Poisson's ratio; however this change to the unit cell decreases it further.

Vat polymerization is an AM method that selectively cures a liquid photopolymer using UV light at extremely fine resolutions such as 50 μm. This process allows mixing ceramic powder into the liquid to create a composite material [45]. Acoustic

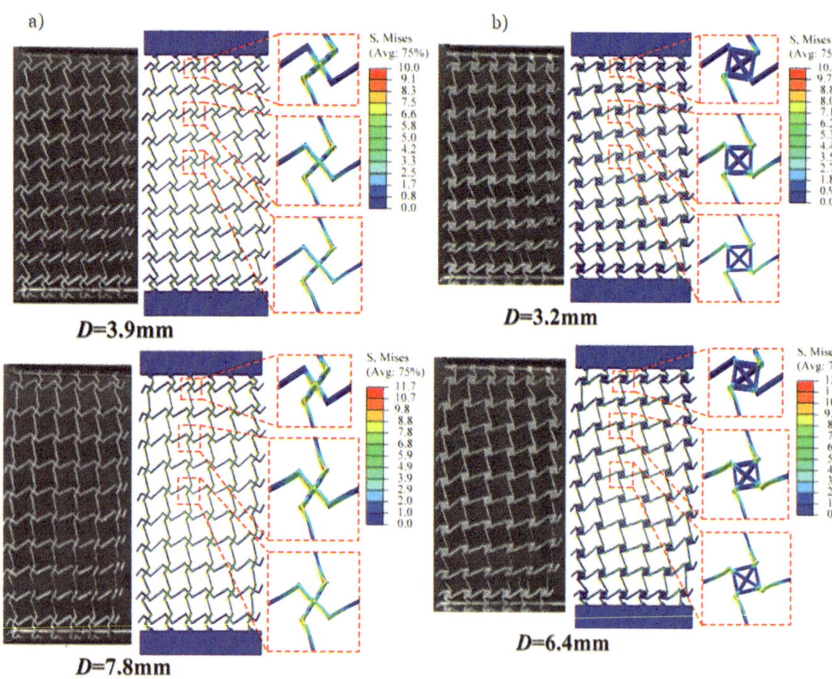

Fig. 5.5 FEA analysis and tension test of material extrusion formed chiral lattice structure at different applied displacements: (**a**) unreinforced lattice and (**b**) reinforced lattice adapted from [41]

metamaterials take advantage of these fine resolutions to create the labyrinth structure necessary to capture and channel acoustic waves [46, 47]. For certain materials, custom manufacturing methods are needed to fabricate MM in high quality. A modified form of vat polymerization was developed to create MMs. This modification allows the operator to tune the thermal expansion of the material during fabrication. By manipulating the trajectory of fibers with a magnetic wiper nickel-coated carbon fibers can be oriented for enhanced structural and thermal performance inside 3d printed structures [48]. This method allows for the creation of composite 3D prints with tunable localized thermal expansion capabilities. Figure 5.6 demonstrates that as a magnet is passed over the nickel-coated carbon fibers, their poles align in response to the magnet as that magnet is driven in a particular direction. This is taken further by being able to pass the magnet in multiple directions over the material allowing for the nickel-coated carbon fibers to align in multiple local orientations creating a local thermal expansion zone across a single structure.

Powder bed fusion (PBF) is a form of AM that sinters or melts a thin layer of powders to form the layers of the material. This method primarily uses metal and polymer feedstock; however, there are some ceramic processes which make use of UV curing binders to produce composite ceramic or pure ceramic parts [49, 50]. Many energy absorbing and auxetic materials are developed using PBF, as this manufacturing method allows for higher grades of material to be printed and tested for use in a wider range of applications than polymers on their own [51]. Figure 5.7 provides an example of PBF manufactured aluminum pantographic fabrics undergoing a torsion test with the powder used to fabricate the samples on the right [50].

A study looked into the PBF method to create an acoustic MM utilizing the principles of the Fabry-Perot resonator to channel sound through the material [52]. Testing in both nylon and aluminum multiple parameters such as corner fileting, flaring edges, and surface roughness were assessed to determine the ability for the material to produce the desired resonance effect. The experiments showed viability of operating conditions between 200 and 800 kHz frequency range in an underwater environment.

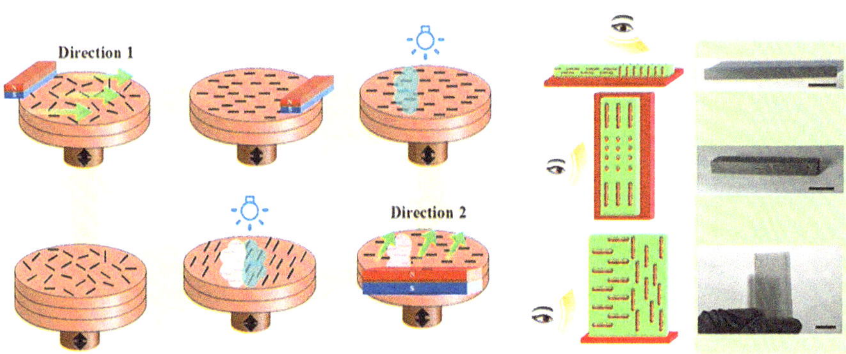

Fig. 5.6 Magnetic wiper to inform the position of the NiCfs and resulting localized heat conduction across multiple trials and orientations adapted from [45]

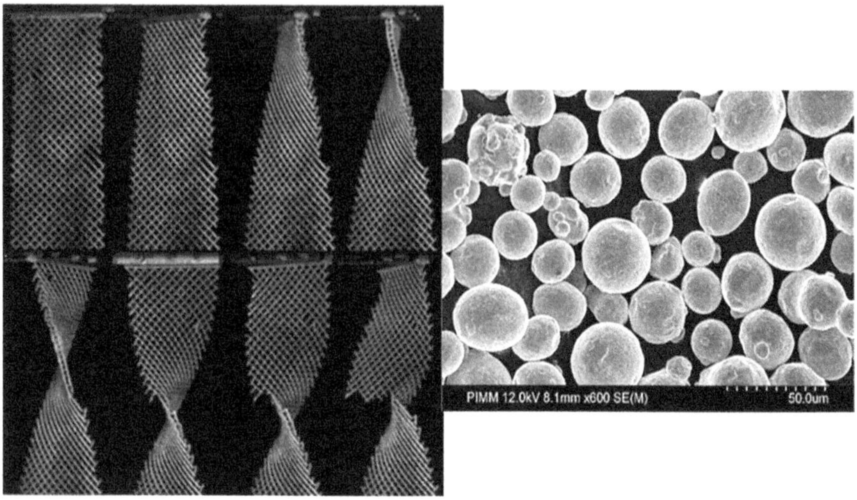

Fig. 5.7 Torsion test of aluminum PBF structure with powder microstructure adapted from [47]

5.5 Traditional Manufacturing Methods

Traditional manufacturing refers to processes such as computer numerical control machining, laser cutting, water jetting, and other typical subtractive manufacturing processes used commonly in industry. Formative manufacturing methods such as casting, molding, forging and others are also in the traditional manufacturing category. MMs use simplified 2D models for evaluating the properties of their materials when they are manufactured by traditional methods. For those materials 2.5D manufacturing processes such as laser cutting, and water jetting are often utilized as they can accurately and efficiently cut the lattice for testing [53, 54]. Another unique study used traditional manufacturing techniques to create and validate an energy absorbing acoustic MM [55]. This material was created from a blend of acrylonitrile butadiene styrene and PMMA through thermoforming the unit cells and combining them into the lattice together through ultrasonic welding. A sound absorbing acoustic MM was fabricated by cutting wedge shapes from melamine foam and adhering them in a desired periodic formation to a sheet of 0.1 mm thick aluminum [56]. For a MM requiring large uniform structures such as an impact-resistant armor panel, laser cutting rapidly produces complex patterns from sheets of metal or polymer.

While traditional methods excel in speed and material versatility for larger structures, they often lack the resolution and capability to create the micro, and nanoscale features possible with AM and electrochemical fabrication. However, their continued relevance is assured by their efficiency, scalability, and lower cost for specific applications, highlighting the importance of choosing the right manufacturing technique based on the specific requirements of the material being produced.

Examples of metamaterials include origami folding patterns to create auxetic materials from metals, typically through compression molding of thin sheets

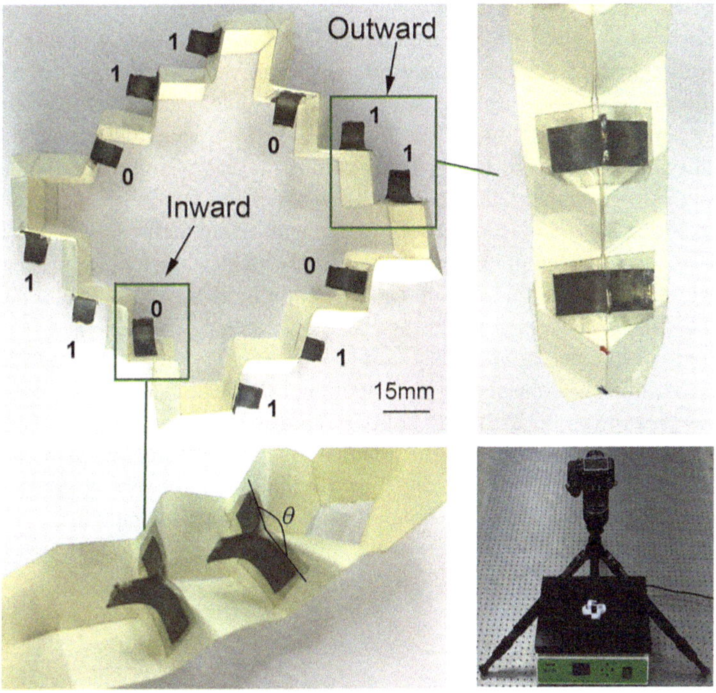

Fig. 5.8 Paper and aluminum foil folded meta-structure with preprogrammed thermal expansion properties adapted from [54]

between a die and a mold. A highly unique origami mechanical MM was created using traditional Japanese techniques to form an auxetic material from paper and metal foil [57]. The purpose behind this structure was to create a thermally tunable auxetic material from the conflicting thermal expansion properties of the two materials. Figure 5.8 provides an illustration for this origami structure where the folded paper creates a chain with the aluminum foil patterned inside and outside the chain, the positions of which dictate the direction of expansion for the unit cell.

5.6 Summary

The link between design and fabrication is critical for metamaterials. Topology optimization algorithms have become the most powerful methods of producing microarchitectures of extraordinarily complex geometries for metamaterials. By iteratively refining a material's form based on predetermined properties, structures that are capable of manipulating light, sound, heat, and force can be produced. Electrochemical fabrication methods have played a vital role in defining the creation of metamaterials. These methods offer precision at the microscale making it ideal for creating electronic components whose geometries need to be at the largest one quarter of the size of the wavelength for the stimulus they will be interacting with. Additive manufacturing has enabled producing complex geometries at the miso and higher scales with precision. AM techniques such as material extrusion, vat polymerization, and powder bed fusion allow for a wide and versatile range of material compositions and finished resolutions making it a popular choice for the fabrication of metamaterials. Despite the prevalence of these new manufacturing methods, there is no replacement for more conventional methods such as computer numerical control machining, laser cutting, and compression molding. These manufacturing methods remain crucial for the creation of metamaterials where ECF and AM do not make sense to form the necessary geometry.

The advancements in both design and fabrication are the keys to creating MMs with further improved properties and functionalities. The future landscape of MMs promises breakthroughs in diverse fields using metamaterials to manipulate electromagnetic waves in medicine and communications, to light-weight energy-absorbing materials for aerospace applications. By continuing to explore the intersection of design and fabrication the full potential of metamaterials can be unlocked.

References

1. Wang, C., Huang, Z., Chen, Z., & Li, Y. (2023). A novel polar mechanical metamaterial with dual deformation characteristics. *International Journal of Mechanical Sciences, 264*, 108827.
2. Yang, F., Gong, J., Yang, E., Guan, Y., He, X., Liu, S., Zhang, X., & Deng, Y. (2019). Ultrabroadband metamaterial absorbers based on ionic liquids. *Applied Physics A, 125*(2), 149.
3. Huang, Y., Guo, W., Jia, J., Wang, L., & Yin, S. (2021). Novel lightweight and protective battery system based on mechanical metamaterials. *Acta Mechanica Solida Sinica, 34*(6), 862–871.
4. Srivastava, H., Pearce, H., Mac, G., & Gupta, N. (2022). Determination of fiber content in 3-D printed composite parts using image analysis. *IEEE Embedded Systems Letters, 14*(3), 115–118.
5. Mac, G., Pearce, H., Karri, R., & Gupta, N. (2021). Uncertainty quantification in dimensions dataset of additive manufactured NIST standard test artifact. *Data in Brief, 38*, 107286.
6. Li, W., Mac, G., Tsoutsos, N. G., Gupta, N., & Karri, R. (2020). Computer aided design (CAD) model search and retrieval using frequency domain file conversion. *Additive Manufacturing, 36*, 101554.
7. Yamaguchi, K., Miyazawa, Y., Yasuda, H., Song, Y., Shimokawa, S., Gandhi, U., & Yang, J. (2023). Post-fabrication tuning of origami-inspired mechanical metamaterials based on Tachi-Miura polyhedron. *Materials & Design, 233*, 112170.
8. Du, Z., Zhang, W., Zhang, Y., Xue, R., & Guo, X. (2019). Structural topology optimization involving bi-modulus materials with asymmetric properties in tension and compression. *Computational Mechanics, 63*(2), 335–363.
9. Anaya-Jaimes, L. M., Vicente, W. M., & Pavanello, R. (2022). Metamaterials design with a desired thermal expansion using a multi-material BESO method. *Structural and Multidisciplinary Optimization, 65*(12), 355.
10. Wei, N., Ye, H., Zhang, X., Li, J., & Sui, Y. (2022). Topology optimization for design of hybrid lattice structures with multiple microstructure configurations. *Acta Mechanica Solida Sinica, 35*(3), 367–383.

11. Viet, N. V., Karathanasopoulos, N., & Zaki, W. (2022). Mechanical attributes and wave propagation characteristics of TPMS lattice structures. *Mechanics of Materials, 172*, 104363.
12. Li, Z., Wang, L., Geng, X., Chen, W., & Han, B. (2024). Complex uncertainty-oriented robust topology optimization for multiple mechanical metamaterials based on double-layer mesh. *Computer Methods in Applied Mechanics and Engineering, 419*, 116589.
13. Brown, N. K., Garland, A. P., Fadel, G. M., & Li, G. (2023). Deep reinforcement learning for the rapid on-demand design of mechanical metamaterials with targeted nonlinear deformation responses. *Engineering Applications of Artificial Intelligence, 126*, 106998.
14. Wu, K., Sigmund, O., & Du, J. (2021). Design of metamaterial mechanisms using robust topology optimization and variable linking scheme. *Structural and Multidisciplinary Optimization, 63*(4), 1975–1988.
15. Liu, Z., Xu, H., & Zhu, P. (2020). An adaptive multi-fidelity approach for design optimization of mesostructure-structure systems. *Structural and Multidisciplinary Optimization, 62*(1), 375–386.
16. Zhang, G., & Khandelwal, K. (2020). Topology optimization of dissipative metamaterials at finite strains based on nonlinear homogenization. *Structural and Multidisciplinary Optimization, 62*(3), 1419–1455.
17. Zhu, Z., Wang, Z., Liu, T., Luo, X., Qiu, C., & Hu, R. (2023). Field-coupling topology design of general transformation multiphysics metamaterials with different functions and arbitrary shapes. *Cell Reports Physical Science, 4*(8), 101540.
18. Abali, B. E., & Barchiesi, E. (2021). Additive manufacturing introduced substructure and computational determination of metamaterials parameters by means of the asymptotic homogenization. *Continuum Mechanics and Thermodynamics, 33*(4), 993–1009.
19. Giorgio, I. (2021). Lattice shells composed of two families of curved Kirchhoff rods: An archetypal example, topology optimization of a cycloidal metamaterial. *Continuum Mechanics and Thermodynamics, 33*(4), 1063–1082.
20. Deng, B., Zareei, A., Ding, X., Weaver, J. C., Rycroft, C. H., & Bertoldi, K. (2022). Inverse design of mechanical metamaterials with target nonlinear response via a neural accelerated evolution strategy. *Advanced Materials, 34*(41), 2206238.
21. Noguchi, Y., Yamada, T., Izui, K., & Nishiwaki, S. (2018). Topology optimization for hyperbolic acoustic metamaterials using a high-frequency homogenization method. *Computer Methods in Applied Mechanics and Engineering, 335*, 419–471.
22. Li, H., Li, H., Xiao, M., Zhang, Y., Fu, J., & Gao, L. (2020). Robust topology optimization of thermoelastic metamaterials considering hybrid uncertainties of material property. *Composite Structures, 248*, 112477.
23. Zhao, S., Zhang, Y., Zhang, Y., Zhang, W., Yang, J., & Kitipornchai, S. (2022). Genetic programming-assisted micromechanical models of graphene origami-enabled metal metamaterials. *Acta Materialia, 228*, 117791.
24. Han, Z., & Wei, K. (2022). Multi-material topology optimization and additive manufacturing for metamaterials incorporating double negative indexes of Poisson's ratio and thermal expansion. *Additive Manufacturing, 54*, 102742.
25. Chen, F., Pinisetty, D., & Gupta, N. (2022). Study of the compressive properties of adhesively bonded carbon fiber laminates at different strain rates. *The Journal of Adhesion, 98*(16), 2582–2598.
26. Poveda, R. L., & Gupta, N. (2013). Post-impact residual high strain rate compressive properties of carbon fiber laminates. *Journal of Reinforced Plastics and Composites, 32*(8), 564–573.
27. Xu, X., & Gupta, N. (2019). Artificial neural network approach to determine elastic modulus of carbon fiber-reinforced laminates. *JOM, 71*(11), 4015–4023.
28. Jia, Z., Hao, S., Wen, J., Li, S., Peng, W., Huang, R., & Xu, X. (2020). Electrochemical fabrication of metal–organic frameworks membranes and films: A review. *Microporous and Mesoporous Materials, 305*, 110322.

29. Long, L., Taylor, S., Ying, X., & Wang, L. (2019). Thermally-switchable spectrally-selective infrared metamaterial absorber/emitter by tuning magnetic polariton with a phase-change VO2 layer. *Materials Today Energy, 13*, 214–220.
30. Du, X., Yan, F., Wang, W., Zhang, L., Bai, Z., Zhou, H., & Hou, Y. (2021). Thermally-stable graphene metamaterial absorber with excellent tunability for high-performance refractive index sensing in the terahertz band. *Optics & Laser Technology, 144*, 107409.
31. Lee, N., Lim, J.-S., Chang, I., Lee, D., & Cho, H. H. (2021). Transparent metamaterials for multispectral camouflage with thermal management. *International Journal of Heat and Mass Transfer, 173*, 121173.
32. Akhtar, M. N., Makhdoom, S., Baqir, M. A., Yousaf, M., Khan, M. A., Batoo, K. M., & Hussain, S. (2022). Enhanced structural, electromagnetic and absorption features of CoSm ferrite-metamaterial absorbers through synergistic effects. *Ceramics International, 48*(20), 29561–29571.
33. Qiao, J., & Liu, Y. (2021). Broadband metamaterial absorber with thermal insensitivity up to 600 °C. *Applied Physics A, 127*(9), 691.
34. Mei, H., Yang, W., Zhao, X., Yao, L., Yao, Y., Chen, C., & Cheng, L. (2021). In-situ growth of SiC nanowires@carbon nanotubes on 3D printed metamaterial structures to enhance electromagnetic wave absorption. *Materials & Design, 197*, 109271.
35. Pan, W., Yan, Y., Ma, Y., & Shen, D. (2019). A terahertz metamaterial based on electromagnetically induced transparency effect and its sensing performance. *Optics Communications, 431*, 115–119.
36. Ma, M., Lian, X., Tian, J., & Yang, R. (2023). A dynamically switchable metamaterial device with multiple functions based on vanadium dioxide. *Photonics and Nanostructures—Fundamentals and Applications, 57*, 101197.
37. Zhang, J., Li, Z., Shao, L., Xiao, F., & Zhu, W. (2021). Active modulation of electromagnetically induced transparency analog in graphene-based microwave metamaterial. *Carbon, 183*, 850–857.
38. Fan, J., Zhang, L., Wei, S., Zhang, Z., Choi, S.-K., Song, B., & Shi, Y. (2021). A review of additive manufacturing of metamaterials and developing trends. *Materials Today, 50*, 303–328.
39. Montgomery, S. M., Kuang, X., Armstrong, C. D., & Qi, H. J. (2020). Recent advances in additive manufacturing of active mechanical metamaterials. *Current Opinion in Solid State and Materials Science, 24*(5), 100869.
40. Zhou, X., Ren, L., Song, Z., Li, G., Zhang, J., Li, B., Wu, Q., Li, W., Ren, L., & Liu, Q. (2023). Advances in 3D/4D printing of mechanical metamaterials: From manufacturing to applications. *Composites Part B: Engineering, 254*, 110585.
41. Peng, Y.-G., Shen, Y.-X., Geng, Z.-G., Li, P.-Q., Zhu, J., & Zhu, X.-F. (2020). Super-resolution acoustic image montage via a biaxial metamaterial lens. *Science Bulletin, 65*(12), 1022–1029.
42. Liu, C. R., Wu, J. H., Lu, K., Zhao, Z. T., & Huang, Z. (2019). Acoustical siphon effect for reducing the thickness in membrane-type metamaterials with low-frequency broadband absorption. *Applied Acoustics, 148*, 1–8.
43. Zeng, Y., Cao, L., Wan, S., Guo, T., Wang, Y.-F., Du, Q.-J., Assouar, B., & Wang, Y.-S. (2022). Seismic metamaterials: Generating low-frequency bandgaps induced by inertial amplification. *International Journal of Mechanical Sciences, 221*, 107224.
44. Zhu, Y., Jiang, S., Zhang, Q., Li, J., Yu, C., & Zhang, C. (2022). A novel monoclinic auxetic metamaterial with tunable mechanical properties. *International Journal of Mechanical Sciences, 236*, 107750.
45. Jeong, H., Cui, Y., Tentzeris, M. M., & Lim, S. (2020). Hybrid (3D and inkjet) printed electromagnetic pressure sensor using metamaterial absorber. *Additive Manufacturing, 35*, 101405.
46. Zhang, Z., Scarpa, F., Bednarcyk, B. A., & Chen, Y. (2021). Harnessing fractal cuts to design robust lattice metamaterials for energy dissipation. *Additive Manufacturing, 46*, 102126.
47. Xiang, X., Wu, X., Li, X., Wu, P., He, H., Mu, Q., Wang, S., Huang, Y., & Wen, W. (2020). Ultra-open ventilated metamaterial absorbers for sound-silencing applications in environment with free air flows. *Extreme Mechanics Letters, 39*, 100786.

48. Ren, L., Zhou, X., Xue, J., Song, Z., Li, B., Liu, Q., & Zhao, C. (2019). Thermal metamaterials with site-specific thermal properties fabricated by 3D magnetic printing. *Advanced Materials Technologies, 4*(7), 1900296.
49. Lan, X., Yang, S., Dong, Y., Wang, Z., & Li, H. (2023). Mechanical model of strut-based metamaterials based on direct stiffness method. *International Journal of Mechanical Sciences, 263*, 108791.
50. Ciallella, A. (2020). Research perspective on multiphysics and multiscale materials: A paradigmatic case. *Continuum Mechanics and Thermodynamics, 32*(3), 527–539.
51. Xu, W., Zhang, L., Zhang, B., Zhang, H., Liu, Z., & Zhu, P. (2023). Crushing behavior of contact-aided AlSi10Mg sandwich structure based on chiral mechanical metamaterials. *International Journal of Mechanical Sciences, 260*, 108636.
52. Askari, M., Hutchins, D. A., Watson, R. L., Astolfi, L., Nie, L., Freear, S., Thomas, P. J., Laureti, S., Ricci, M., Clark, M., & Clare, A. T. (2020). An ultrasonic metallic Fabry–Pérot metamaterial for use in water. *Additive Manufacturing, 35*, 101309.
53. Mizzi, L., Salvati, E., Spaggiari, A., Tan, J.-C., & Korsunsky, A. M. (2020). Highly stretchable two-dimensional auxetic metamaterial sheets fabricated via direct-laser cutting. *International Journal of Mechanical Sciences, 167*, 105242.
54. Cheng, X., Zhang, Y., Ren, X., Han, D., Jiang, W., Zhang, X. G., Luo, H. C., & Xie, Y. M. (2022). Design and mechanical characteristics of auxetic metamaterial with tunable stiffness. *International Journal of Mechanical Sciences, 223*, 107286.
55. de Melo Filho, N. G. R., Claeys, C., Deckers, E., & Desmet, W. (2019). Realisation of a thermoformed vibro-acoustic metamaterial for increased STL in acoustic resonance driven environments. *Applied Acoustics, 156*, 78–82.
56. Ji, G., Fang, Y., & Zhou, J. (2020). Porous acoustic metamaterials in an inverted wedge shape. *Extreme Mechanics Letters, 36*, 100648.
57. Yang, N., Zhang, M., & Zhu, R. (2020). 3D kirigami metamaterials with coded thermal expansion properties. *Extreme Mechanics Letters, 40*, 100912.

Chapter 6
Current Limitations and Future Directions

List of Symbols

C_c	Coupling tensor
C_e	Elasticity tensor
C_{micro}	Elasticity tensor
$\nabla \boldsymbol{u}$	Kinetic energy
$\mathbb{J}_c$	Coupling tensor
$\mathbb{J}_{\text{micro}}$	Micro inertia density tensor
$\mathbb{T}$	Gradient micro inertia density tensor
$\mathbb{T}_c$	Coupling tensor
$\mathbb{L}$	Elasticity tensor
$\mathbb{L}_c$	Coupling tensor

6.1 Introduction

The design and functionality of metamaterials (MMs) are inherently influenced by the medium and stimuli they interact with, presenting unique challenges. One such challenge arises from the directional nature of waves interactions, where signals are often transmitted and reflected in spectrum patterns. In scenarios of unidirectional transmission, precise reflection angles are required. However, for a broader range of applications, the anisotropic nature of metamaterials, like considerations carpenters make for the grain orientation in timber construction, become critical factors in the design process. Incorrect orientation can lead to suboptimal functionality or outright failure of the material in operation [1, 2].

Furthermore, metamaterials exhibit a susceptibility to cyclic loading and fatigue, which can compromise their integrity. Repeated exposure to varying rates of force and thermal stress has highlighted a flaw revealing the brittleness inherent in many MM designs, significantly reducing their lifespan [3, 4]. Additionally, the specificity of materials attenuation to certain frequency bands presents another layer of

N. Gupta, C. Beckwith, *Architected Metamaterials: Design Principles and Properties*, https://doi.org/10.1007/978-3-031-98330-6_6

complexity. Designed to interact within specific frequency ranges, these materials can introduce interference or "noise" when integrated with other materials of differing natural frequencies, potentially compromising the overall system's function [5, 6]. As an illustration of these challenges, Fig. 6.1 provides a depiction of some of these issues that arise in developing metamaterials: (a) illustrates the inaccuracies between simulations and two printed models as well as the testing results of each [7] and (b) highlights inaccuracies that arise in acoustic MM testing for sound absorption against simulated results [8]. Figure 6.2 continues this trend and highlights other examples of issues that can arise in the development of metamaterials. Figure 6.2a presents an example of a cloaking/transparency switchable thermal MM demonstrated to be susceptible to thermal cycling [9] and (b) provides anisotropic

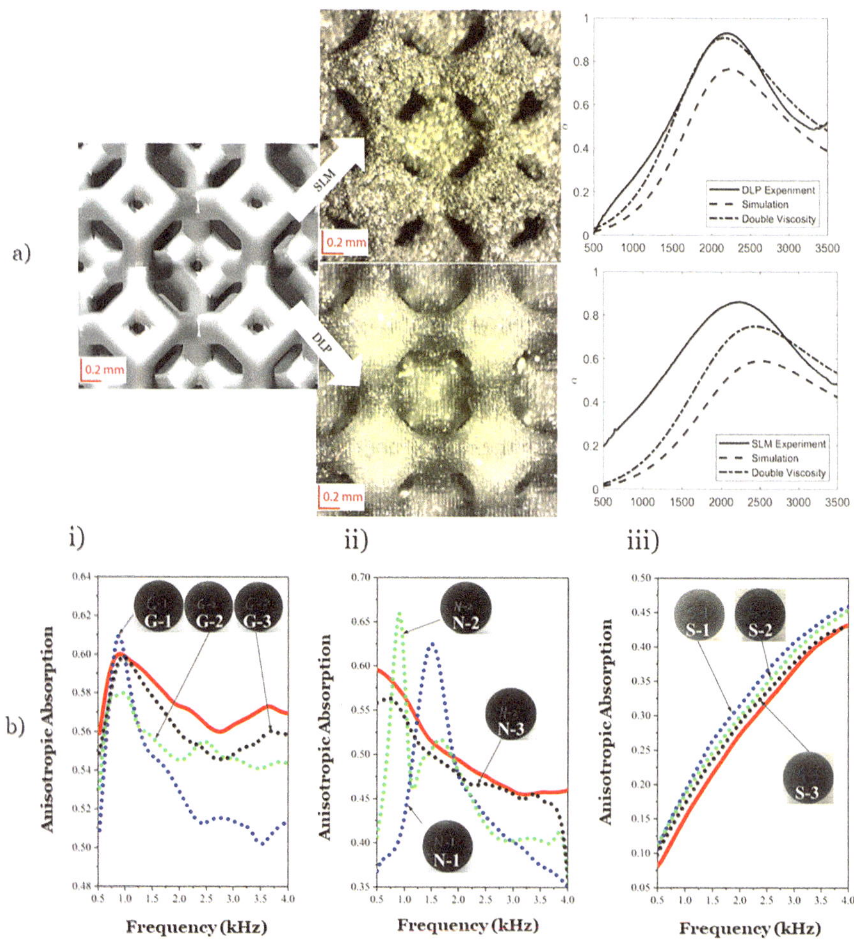

Fig. 6.1 Examples of developmental limitations of MMs. (**a**) Simulation against PBF print and VP print with testing results adapted from [7]. (**b**) Absorption over frequency simulations against validated models of acoustic MM adapted from [8]

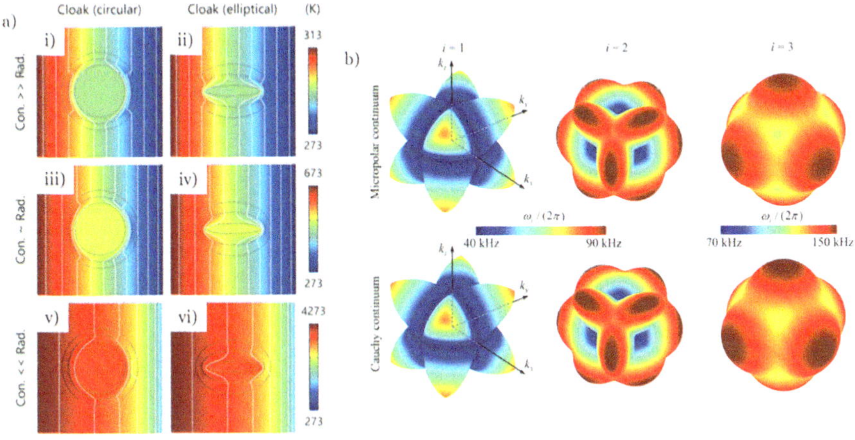

Fig. 6.2 Examples of developmental limitations of MMs. (**a**) Unstable thermal cloak adapted from [9]. (**b**) Anisotropy analysis for mechanical MM adapted from [10]

analysis of an auxetic mechanical MM unit cell illustrating the complexities in understanding the anisotropic behaviors [10].

6.2 Anisotropic by Design

Anisotropy, the characteristic of a material whose properties are direction dependent, is central to the design and functionality of metamaterials. Unlike traditional materials, such as metals, and plastics, which typically exhibit isotropic behavior, uniform properties in all directions, due to the nature of these materials being engineered to perform highly niche tasks, they often demonstrate varied responses when subjected to stimuli from different orientations. This directional variance is notably significant at the mesoscale where the introduction of specific configurations induces anisotropic characteristics due to the comparatively large scale of the lattice [11, 12].

The intricate designs of MMs inherently embrace anisotropy, as their unique properties are often a direct result of their structured geometry [13, 14]. This is especially true for MMs with engineered tasks such as energy absorption, where the effectiveness of the lattice structure is contingent upon its orientation relative to the incident stimuli, such as a solar absorber. In dynamic systems, this orientation dependence introduces complexity in predicting the most efficient energy absorption conditions, with axial or planar designs faltering under oblique loads.

To navigate the challenges posed by anisotropy, one approach is to analyze the impact of lattice orientation and consider transitioning from 2D to more complex 3D unit cells [15, 16]. While this does not eliminate anisotropy, it broadens the design spectrum, allowing for configurations to exploit the inherent strengths of the materials. For instance, Fig. 6.3 details the testing processes for evaluating the anisotropy of a metamaterial [16]. For this case, a face-centered cubic lattice was rotated incrementally from 0° to 45° and compressed to determine the mode shapes and determine the optimal loading orientation such that the stress-strain curve is

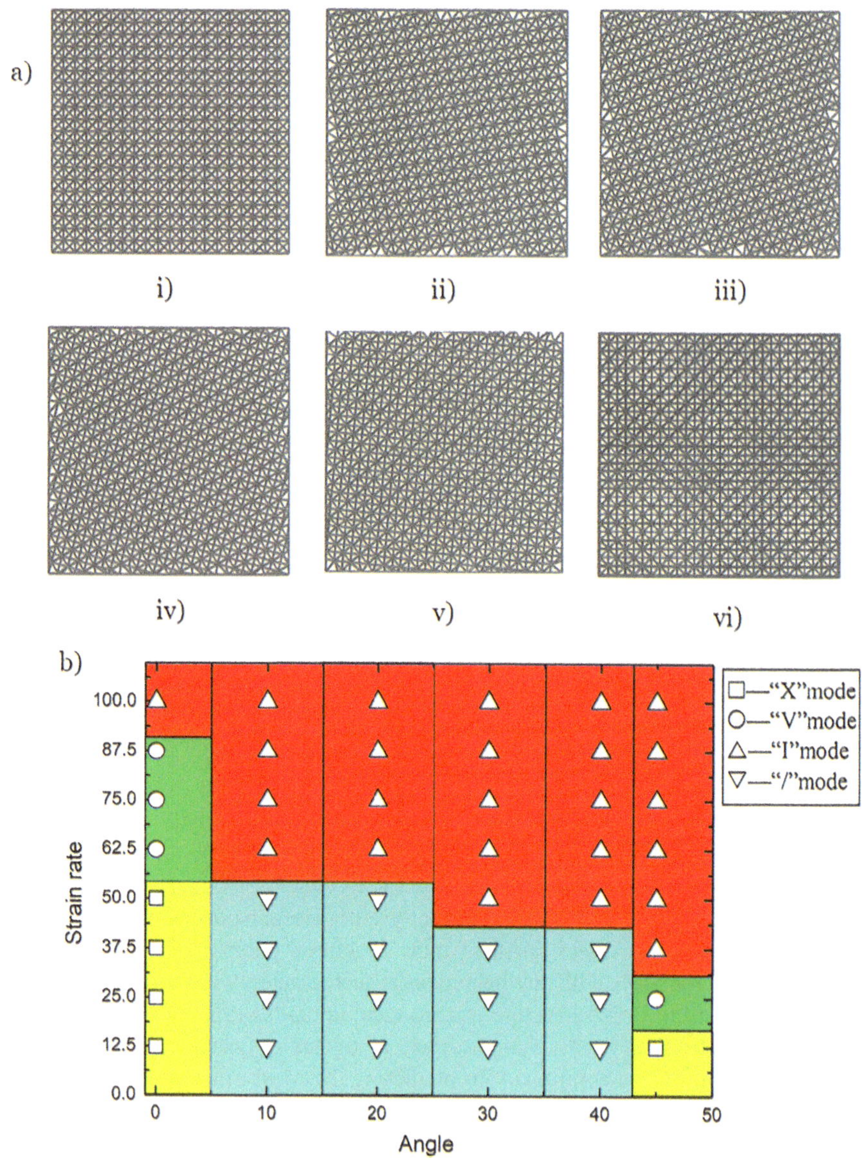

Fig. 6.3 (**a**) Anisotropic analysis of mechanical absorbing metamaterial through compression testing of face-centered cubic lattice at multiple angles: (i) = 0°, (ii) = 10°, (iii) = 20°, (iv) = 30°, (v) = 40°, (iv) = 45°. (**b**) Mode shape distributions observed resulting from the compression of the materials at different strain rates adapted from [16]

optimized. Figure 6.4 provides the stress-strain curves for these different angles. It can be observed from this that the lowest angle and highest angle (Fig. 6.4a, d) result in the highest stress response whereas the off angles show lower mechanical properties (Fig. 6.4b, c).

Anisotropy plays an important role in the behavior of acoustic MM [17]. Studies used the relaxed micromorphic model to describe the mechanical behavior by introducing fourth-order elasticity tensors demonstrating the effects of anisotropy are present in both the classical analysis through Bloch-Floquet and within the relaxed model. The relaxed micromorphic model is a generalized continuum theory used to describe materials with complex microstructures by introducing additional degrees of freedom. This is a reduction of the full micromorphic model by reducing the number of constitutive parameters but still capturing size effects and nonlocal behaviors and is applied as [17]:

$$\begin{aligned}\rho \boldsymbol{u}_{,tt} - \boldsymbol{Div}\left(\mathbb{T}\ \boldsymbol{sym}\nabla \boldsymbol{u}_{,tt}\right) - \boldsymbol{Div}\left(\mathbb{T}_c\ \boldsymbol{sym}\nabla \boldsymbol{u}_{,tt}\right) \\ = \boldsymbol{Div}\left(\mathbb{C}_e \boldsymbol{sym}\left(\nabla \boldsymbol{u} - \boldsymbol{P}\right) + \mathbb{C}_c\ \boldsymbol{skew}\left(\nabla \boldsymbol{u} - \boldsymbol{P}\right)\right),\end{aligned} \tag{6.1}$$

$$\begin{aligned}\mathbb{J}_{\text{micro}} sym\text{P}_{,tt} = \mathbb{C}_e sym\left(\nabla \text{u} - \text{P}\right) - \mathbb{C}_{\text{micro}}\text{symP} \\ -\mu L^2 \text{symCurl}\left(\mathbb{L}\text{symCurlP} + \mathbb{L}_c \text{skewCurlP}\right),\end{aligned} \tag{6.2}$$

$$\begin{aligned}\mathbb{J}_c \text{skewP}_{,tt} = \mathbb{C}_c \text{skew}\left(\nabla \text{u} - \text{P}\right) \\ -\mu L_c^2 \text{symCurl}\left(\mathbb{L}\text{symCurlP} + \mathbb{L}_c \text{skewCurlP}\right)\end{aligned} \tag{6.3}$$

This analysis has shown anisotropic band-gap behavior for an acoustic MM with tetragonal lattice structure. This level of analysis is important to demonstrate the effects of anisotropy in a metamaterial which is not always present in classical methods.

Overcoming the anisotropic qualities of the lattice structure and altering the material such that it is optimized for its loading conditions is the first step in expanding MMs beyond the laboratory setting. The material's functionality is chosen first and the geometry often results from that in the form of boundary conditions. Understanding the limitations these boundary conditions impose allows for greater evolution in the material's utility.

6.3 Establishing a Robust Testing Suite for Metamaterial Performance

Producing metamaterials can be costly and time consuming, therefore developing testing methods to evaluate multiple properties in a single test is a crucial factor in determining the usefulness of metamaterials outside of the laboratory. There is a lack of data showing the ultimate life expectancy of metamaterials beyond

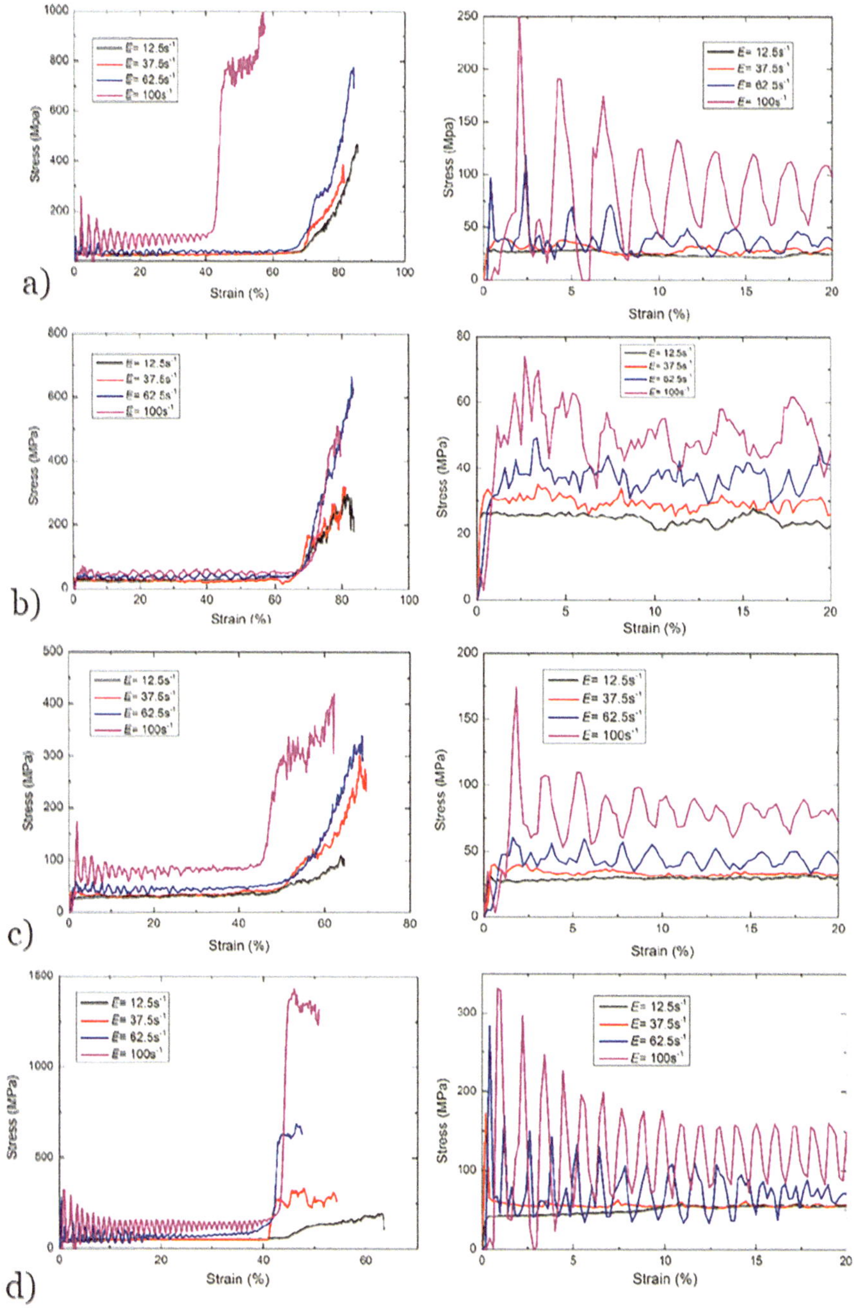

Fig. 6.4 Distribution classification delineating between "X," "V," "I," "/" representing mode shapes observed during deformation. Right image set depicts the stress-strain curves of the lattices at orientations (**a**) = 0°, (**b**) = 10°, (**c**) = 30°, and (**d**) = 45° adapted from [16]

determining their application relevant properties. Ways to combat this limitation include performing dynamic mechanical analysis to establish a master curve of the material's properties across a spectrum of temperatures and frequencies, weather testing to understand the degradation pattern for the material's use in outdoor conditions, and high strain rate testing to show the energy absorption and failure mechanisms of the material for protective applications [18–20]. Utilizing digital image correlation during mechanical testing can be used to establish the training data necessary to train a machine learning model to predict the mechanical properties of future metamaterials without the need for many tests [21, 22]. Combining this with forms of nondestructive evaluation for ensuring material performance will allow research into these materials to accelerate further by gathering more data about the material without fabricating more samples. In many cases for the additively manufactured samples, nondestructive testing can be deployed during fabrication to construct a preliminary set of material properties [23], establish the thermal history of the material, and characterize any defects that appear in the sample during manufacturing which could lead to a degradation in the material's properties during testing [24, 25].

Carbon nanofibers (CNFs) are a common filler material added to composites and metamaterials to increase mechanical properties. The effects of these fibers are widely studied, and composites composed of these materials are compared to one another to establish a consensus on the performance of these materials under differing compositions [26–28]. Tensile testing, weather testing, and compression testing have all been performed on these materials ad nauseum. For metamaterials to reach this same level of acceptance in wider applications they must eventually reach a consensus of how these materials will perform over their lifetime. High strain rate and impact tests are necessary for structures that are exposed to such environments, for example, automotive and aerospace structures must absorb energy efficiently and deform predictably under high strain rates to ensure safety and functionality [29, 30].

A novel approach taken to establish the performance of these materials compared the quasi-static and high strain rate compressive strength of syntactic foams created using varying weight percentages of carbon nanofiber and of hollow microglass spheres. The material properties depend on the properties of the ingredients and their volume fractions; however, without comprehensive testing the materials properties are not comparable. Figure 6.5 depicts the variety of combinations of foams tested along with some notable outcomes from this. Figure 6.5b demonstrates a deformation that occurs in the s under quasi-static loading, and in Fig. 6.5c the results of the compression testing become visible [31]. There is a large gap in the distribution of the compression strength under quasi-static strain, where the high strain rate is more concentrated. These effects would not be shown if the comparison charts are not developed based on a wide range of material compositions and loading conditions.

Combining these mechanical tests creates a comprehensive profile of the syntactic foams' performance. This holistic approach to testing enables engineers to identify strengths and limitations, leading to the development of advanced syntactic

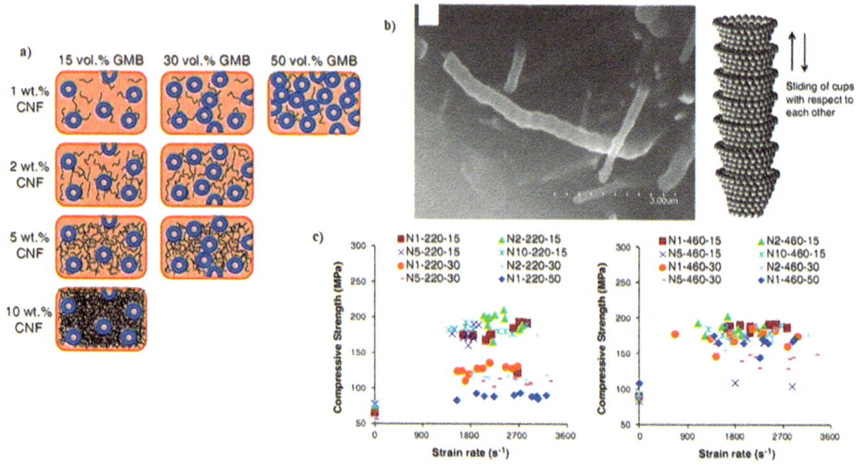

Fig. 6.5 (**a**) Schematic representation of -reinforced syntactic foams made with hollow glass microspheres. (**b**) SEM image of carbon nanofibers deformed after being subjected to quasi-static compression, with a schematic of the sliding effect which occurred on the layers of deformed graphite layers. (**c**) Comparison of quasi-static and high strain rate compressive values for carbon nanofiber syntactic foams adapted from [31]

foams tailored for specific applications. Integrating results from diverse testing methods allows for the creation of robust, durable, and high-performance materials, ensuring they meet the demands of their intended use across various industrial applications. As metamaterials continue to develop, this holistic approach of combining testing methods will be required to build a more robust argument in probing the necessity of these materials over the current materials used in industry.

6.4 Bridging Theory and Practice in Metamaterial Design

The development of MMs is driven by precise theoretical models and boundary conditions tailored to specific types of incident stimuli. When interacting with waves it is quite common for the unit cell size of an MM to be smaller than one quarter of the wavelength of the incident wave [32]. This guideline ensures the effective manipulation of waves, both electromagnetic and acoustic, by the MMs structure. However, the stringent nature of this criteria often confines the form of these materials [33]. The highly intricate nature of these unit cells coupled with their small size when interacting with these waves leaves little room for error in the manufacturing process. Figure 6.1a, b illustrates two examples of this where deviations in the manufacturing process are theorized to be the source of the discrepancies in the performance of the material.

Variations in analysis and laboratory conditions, such as differences in simulation software, computational resources, and experimental set-ups, can significantly impact the predicted performance of MMs. This deviation makes evaluating the performance of the material difficult, especially when strict wavelength targets are being matched [34]. Moreover, the fabrication and practical testing can reveal deviations from theoretical predictions, especially when the operational wavelengths stray from the designed spectrum [35]. Such deviations are critical for MMs designed to manipulate waves, or deal with vibrations. For instance, exposure to vibrations at or near a material's natural frequency can compromise the structural integrity. Similarly, electromagnetic MMs optimized for specific band regions may fail to transmit or emit noise when subjected to signals outside of their designed frequency range [4].

6.5 Future Directions in Metamaterials Research

As metamaterials continue to captivate the scientific community, the volume of research underscores the field's vibrancy and potential. Each year, a myriad of studies unveils novel MMs, pushing the boundaries of what is theoretically possible and exploring an ever-expanding array of applications. Yet, the translation of these advanced materials from the laboratory setting to real-world applications remains a formidable challenge. The primary obstacle lies in overcoming the failure modes intrinsic to these materials and enhancing their design to ensure reliable performance under practical conditions. The quest to bridge this gap has led researchers to explore innovative fabrication techniques, with additive manufacturing emerging as a particularly promising method. AM offers a rapid and flexible means of manufacturing MMs, allowing for swift iterations and experimental validation. By evaluating these materials in the real world, researchers can gain invaluable insights into their performance, identifying strengths to be leveraged and weaknesses to be addressed before using that same information to iterate and improve the design of the material's digital and theoretical twins.

Moreover, the integration of machine learning into the MM design process represents a transformative shift in how these materials are conceived and optimized. By harnessing vast datasets derived from experimental validations, machine learning models can be trained to predict the performance of MMs designs under a variety of conditions. This predictive capability can significantly streamline the design phase, enabling the use of topology optimization algorithms to refine MMs structure efficiently.

The future of metamaterials is poised at the intersection of advanced manufacturing, computational design, and artificial intelligence. These interdisciplinary approaches hold the key to unlocking the full potential of MMs, transitioning them from laboratory curiosities to integral components of next-generation technologies. The vision is for MMs to evolve into versatile designer materials that engineers can tailor to specific applications with unprecedented precision and efficiency. Digital manufacturing and additive manufacturing are enabling new frontiers in the design and fabrication of metamaterials, offering unprecedented freedom in geometry, material combinations, and functional integration. However, as these technologies become increasingly interconnected and reliant on cyber-physical systems, they also introduce new cybersecurity vulnerabilities. Attack vectors in the digital manufacturing workflow—including CAD file tampering, STL file interception, printer firmware manipulation, and unauthorized access to cloud-based manufacturing platforms—pose significant threats to the integrity, confidentiality, and authenticity of metamaterial designs. These cyber threats can manifest in various forms, such as intentional alteration of internal lattice structures, degradation of material performance, or subtle dimensional changes that compromise the functional properties of the final product. Such attacks are particularly concerning in high-stakes industries like aerospace, defense, and medical devices, where metamaterials are being explored for their lightweight, tunable, and multifunctional properties. Malicious interventions during the design or fabrication stages could lead to catastrophic failure in operation or create undetectable backdoors for exploitation. To address these emerging risks, novel methods of securing the entire digital manufacturing pipeline must be developed. This includes the integration of cryptographic techniques for file authentication, blockchain-based traceability for design provenance, anomaly detection in real-time printing processes, and hardware-level security features within printers and post-processing equipment. Building resilience in both software and hardware systems will be critical to safeguarding the future of metamaterial innovation and ensuring the reliability of components produced through digital manufacturing [36–40].

6.6 Summary

Metamaterials promise to transform various industries by enabling unprecedented manipulations of physical phenomena. However, significant challenges such as material anisotropy and fatigue resistance remain, as well as discrepancies between theoretical predictions and experimental validations. These discrepancies can limit the wider adoption of MMs, as the inconsistency in data and reproducibility hinder their practical application outside of controlled laboratory experiments. Despite these obstacles, the pathway to realizing the full potential of metamaterials is being illuminated by innovative solutions and relentless research efforts. The agility of AM stands out as a crucial enabler for MMs, offering rapid prototyping capabilities that significantly accelerate the cycle of design, testing, and refinement. This approach is vital not only for validating theoretical models but also unveiling new insights through empirical evaluation. In parallel, the integration of machine learning into MM design is paving the way for predictive modeling, enabling the anticipation of material behaviors before physical fabrication. This constructive interaction between computational predictions and experimental validation is key to streamlining the MM development process.

As the field progresses, these technological advancements are increasingly bridging the gap between the envisioned capabilities of MMs and their tangible applications. The fusion of innovative manufacturing techniques, advanced computational tools, and interdisciplinary research is crafting a future where MMs are not just laboratory curiosities but integral components of technological innovation. This evolution points toward a future where MMs contribute to various sectors including telecommunications, aerospace, medicine, and more, transforming theoretical science into practical impactful technologies.

References

1. Bhuwal, A. S., Liu, T., Ashcroft, I., & Sun, W. (2021). Localization and coalescence of imperfect planar FCC truss lattice metamaterials under multiaxial loadings. *Mechanics of Materials, 160*, 103996.
2. Chen, Y., Ye, L., Zhang, Y. X., & Fu, K. (2021). Compression behaviours of 3D-printed CF/PA metamaterials: Experiment and modelling. *International Journal of Mechanical Sciences, 206*, 106634.
3. Ling, F., Zhong, Z., Huang, R., & Zhang, B. (2019). Dynamic-shift single- and double-negative refractive index in a novel three-dimensional metamaterial. *Plasmonics, 14*(3), 695–701.
4. Chirumamilla, M., Krishnamurthy, G. V., Knopp, K., Krekeler, T., Graf, M., Jalas, D., Ritter, M., Störmer, M., Petrov, A. Y., & Eich, M. (2019). Metamaterial emitter for thermophotovoltaics stable up to 1400 °C. *Scientific Reports, 9*(1), 7241.
5. Lang, T., Yu, Z., Zhang, J., Hong, Z., Liu, J., & Wang, P. (2023). Bovine serum albumin detection based on electromagnetically induced transparency in terahertz metamaterial. *Sensors and Actuators A: Physical, 360*, 114522.
6. Lu, X., Wu, X., Xiang, H., Shen, J., Li, Y., Li, Y., & Wang, X. (2022). Triple tunability of phononic bandgaps for three-dimensional printed hollow sphere lattice metamaterials. *International Journal of Mechanical Sciences, 221*, 107166.
7. Rice, H. J., Kennedy, J., Göransson, P., Dowling, L., & Trimble, D. (2020). Design of a Kelvin cell acoustic metamaterial. *Journal of Sound and Vibration, 472*, 115167.
8. Gao, N., & Lu, K. (2020). An underwater metamaterial for broadband acoustic absorption at low frequency. *Applied Acoustics, 169*, 107500.
9. Xu, L., & Huang, J. (2019). Metamaterials for manipulating thermal radiation: Transparency, cloak, and expander. *Physical Review Applied, 12*(4), 044048.
10. Chen, Y., Frenzel, T., Guenneau, S., Kadic, M., & Wegener, M. (2020). Mapping acoustical activity in 3D chiral mechanical metamaterials onto micropolar continuum elasticity. *Journal of the Mechanics and Physics of Solids, 137*, 103877.
11. Qian, J., Gou, P., Pan, H., Zhu, L., Gui, Y. S., Hu, C. M., & An, Z. (2020). Hybrid perfect metamaterial absorber for microwave spin rectification applications. *Scientific Reports, 10*(1), 19240.
12. Gao, Y., & Wang, L. (2022). Nonlocal active metamaterial with feedback control for tunable bandgap and broadband nonreciprocity. *International Journal of Mechanical Sciences, 219*, 107131.
13. Fernández, M., Jamshidian, M., Böhlke, T., Kersting, K., & Weeger, O. (2021). Anisotropic hyperelastic constitutive models for finite deformations combining material theory and data-driven approaches with application to cubic lattice metamaterials. *Computational Mechanics, 67*(2), 653–677.
14. Xu, Y., Hong, Y., Li, M., & He, X. (2023). Underwater low-frequency sound absorption performance and broadband absorption design of membrane-type acoustic metamaterials. *Applied Acoustics, 214*, 109676.

15. Qu, S., Yang, M., Xu, Y., Xiao, S., & Fang, N. X. (2023). Reverberation time control by acoustic metamaterials in a small room. *Building and Environment, 244*, 110753.
16. Wang, P., Bian, Y., Yang, F., Fan, H., & Zheng, B. (2020). Mechanical properties and energy absorption of FCC lattice structures with different orientation angles. *Acta Mechanica, 231*(8), 3129–3144.
17. d'Agostino, M. V., Barbagallo, G., Ghiba, I.-D., Eidel, B., Neff, P., & Madeo, A. (2020). Effective description of anisotropic wave dispersion in mechanical band-gap metamaterials via the relaxed micromorphic model. *Journal of Elasticity, 139*(2), 299–329.
18. Al-Sharab, J., Joshi, B., & Gupta, N. (2012). The effect of multiwall carbon nanotubes (MWCNT) on the high strain rate compressive properties of syntactic nanocomposite foams. *Microscopy and Microanalysis, 18*(S2), 1644–1645.
19. Poveda, R. L., Achar, S., & Gupta, N. (2014). Viscoelastic properties of carbon nanofiber reinforced multiscale syntactic foam. *Composites Part B: Engineering, 58*, 208–216.
20. Zeltmann, S. E., Poveda, R. L., & Gupta, N. (2015). Accelerated environmental degradation and residual flexural analysis of carbon nanofiber reinforced composites. *Polymer Degradation and Stability, 121*, 348–358.
21. Choi, J. Y., Yanamandra, K., Shetty, A., & Gupta, N. (2022). Simultaneous measurement of elastic constants from dynamic mechanical analysis with digital image correlation. *Polymer, 242*, 124562.
22. Choi, J. Y., Yanamandra, K., Shetty, A., & Gupta, N. (2023). Measurement of viscoelastic constants and Poisson's ratio of carbon fiber reinforced composites using in-situ imaging. *Journal of Reinforced Plastics and Composites, 42*(13–14), 638–647.
23. Gupta, N. (2024). Future of non-destructive testing in the era of additive manufacturing and machine learning. *Journal of Non-Destructive Testing, 21*(1), 47–54.
24. AbouelNour, Y., & Gupta, N. (2023). Assisted defect detection by in-process monitoring of additive manufacturing using optical imaging and infrared thermography. *Additive Manufacturing, 67*, 103483.
25. AbouelNour, Y., & Gupta, N. (2024). Comparison of in-situ nondestructive testing and ex-situ methods in additive manufactured specimens for internal feature detection. *Research in Nondestructive Evaluation, 35*(1), 20–31.
26. Dimchev, M., Caeti, R., & Gupta, N. (2010). Effect of carbon nanofibers on tensile and compressive characteristics of hollow particle filled composites. *Materials & Design, 31*(3), 1332–1337.
27. Colloca, M., Gupta, N., & Porfiri, M. (2013). Tensile properties of carbon nanofiber reinforced multiscale syntactic foams. *Composites Part B: Engineering, 44*(1), 584–591.
28. Poveda, R. L., Dorogokupets, G., & Gupta, N. (2013). Carbon nanofiber reinforced syntactic foams: Degradation mechanism for long term moisture exposure and residual compressive properties. *Polymer Degradation and Stability, 98*(10), 2041–2053.
29. Poveda, R. L., & Gupta, N. (2016). Change in failure mode of carbon nanofibers in nanocomposites as a function of loading rate. *Journal of Materials Science, 51*(10), 4917–4927.
30. Gupta, N., Woldesenbet, E., & Mensah, P. (2004). Compression properties of syntactic foams: Effect of cenosphere radius ratio and specimen aspect ratio. *Composites Part A: Applied Science and Manufacturing, 35*(1), 103–111.
31. Poveda, R. L., & Gupta, N. (2014). Carbon-nanofiber-reinforced syntactic foams: Compressive properties and strain rate sensitivity. *JOM, 66*(1), 66–77.
32. Park, C. M., & Lee, S. H. (2019). Zero-reflection acoustic metamaterial with a negative refractive index. *Scientific Reports, 9*(1), 3372.
33. Annamdas, V. G. M., & Soh, C. K. (2018). Application of metamaterial surface Plasmon and waveguide for robotic-arm based structural health monitoring. *Journal of Nondestructive Evaluation, 37*(2), 34.
34. Kraus, J. A., & Rimoli, J. J. (2023). An elastica theory for compressible imperfect beams with application to mechanical metamaterials. *Mechanics Research Communications, 131*, 104147.

35. Ghiyasvand, M., Naser-Moghadasi, M., Lotfi-Neyestanak, A. A., & Nikfarjam, A. (2019). Optimization of the performance of transparent enclosures with deposited gold nanolayers using an analytical model. *Journal of Computational Electronics, 18*(3), 962–970.
36. Chen, F., Mac, G., & Gupta, N. (2017). Security features embedded in computer aided design (CAD) solid models for additive manufacturing. *Materials & Design, 128*, 182–194.
37. Pearce, H., Yanamandra, K., Gupta, N., & Karri, R. (2022). FLAW3D: A Trojan-based cyber attack on the physical outcomes of additive manufacturing. *IEEE/ASME Transactions on Mechatronics, 27*(6), 5361–5370.
38. Beckwith, C., Naicker, H. S., Mehta, S., Udupa, V. R., Nim, N. T., Gadre, V., Pearce, H., Mac, G., & Gupta, N. (2022). Needle in a haystack: Detecting subtle malicious edits to additive manufacturing G-code files. *IEEE Embedded Systems Letters, 14*(3), 111–114.
39. Yanamandra, K., Chen, G. L., Xu, X., Mac, G., & Gupta, N. (2020). Reverse engineering of additive manufactured composite part by toolpath reconstruction using imaging and machine learning. *Composites Science and Technology, 198*, 108318.
40. Mac, G., Mandal, M., Rastogi, A., Gailani, G., Pearce, H., & Gupta, N. (2022). Application of micro-computed tomography for authentication of 3D printed composite parts. *Additive Manufacturing Letters, 3*, 100104.

Index

N. Gupta, C. Beckwith, *Architected Metamaterials: Design Principles and Properties*, https://doi.org/10.1007/978-3-031-98330-6

The manufacturer's authorised representative in the EU is Springer Nature Customer Service Centre GmbH, Europaplatz 3, 69115 Heidelberg, Germany. If you have any concerns regarding our products, please contact ProductSafety@springernature.com

Printed and bound by CPI Group (UK) Ltd, Croydon, CR0 4YY
30/09/2026
02225471-0006